Workshop on New Directions in Mössbauer Spectroscopy (Argonne 1977)

AIP Conference Proceedings
Series Editor: Hugh C. Wolfe
No. 38

Workshop on New Directions in Mössbauer Spectroscopy (Argonne 1977)

Editor

Gilbert J. Perlow
Argonne

American Institute of Physics

New York   1977

L.C. Catalog Card No. 77-90635
ISBN 0-88318-137-1
ERDA CONF- 770688

American Institute of Physics
335 East 45th Street
New York, N.Y. 10017

Printed in the United States of America

INTRODUCTION TO THE PROCEEDINGS

The Workshop on New Directions in Mössbauer Spectroscopy was organized to fill a need which I and a number of people to whom I spoke felt was no longer being met by the biennial international conferences on Applications of the Mössbauer Effect. The uses of Mössbauer spectroscopy for solid state properties, chemistry, and biology are so numerous that one can understand the reluctance of the organizing committees to commit any appreciable time in a major conference to other applications.

The first circular letter was sent to about 40 persons with the request that they propagate it further, selectively. A section from this letter stated:

"The biennial conferences stress: materials science and the properties of solids, chemistry and chemical physics, and biology. They and some peripheral fields (Archeology, etc.) are therefore adequately covered and should not be part of the workshop program.

A personal list of topics followed which was amended and enlarged as a result of correspondence. The final list was:

TOPICS OF THE WORKSHOP

1. New and neglected techniques, including synchrotron radiation sources, focussing and guiding, resonant Bragg scattering, Rayleigh scattering, selective excitation, techniques for detecting small effects, new uses of ultra-low temperatures.
2. Coherence phenomena.
3. New isotopes of special interest.
4. Relativity experiments.
5. Mixed Technology: r-f experiments, β-spectrometry, etc.
6. Nuclear physics, electromagnetic moments, charge and moment distribution in nuclei, symmetries from Mössbauer experiments, experiments in accelerator beams.
7. Other ideas, experiments, or theory that fit within the framework of the Workshop title.

The Workshop was in the nature of an experiment, one of whose aims was to ascertain if there exists a constituency for such subjects. It appears that there is. The attendance was 62 from 12 countries, plus additional unregistered auditors from the local physics community. The dates of the workshop were chosen to permit attendance of the participants at the subsequent conference on Hyperfine interactions at Madison, N.J.

All of the housekeeping tasks that make the management of conferences an interminable succession of details were superbly taken care of by the Argonne Conference Management Office, headed by Miriam L. Holden and assisted by Dorothy Burdzinski and Rose Lorenz. The editorial manipulation of the manuscripts in this volume and the typing and disseminating of the workshop announcements are correspondence were ably performed by Diane Kokosz. Eric Shakin rendered valuable service in recording and transcribing comments, and Gopal Shenoy gave expert advice on getting started. I thank Hugh Wolfe and the management of the American Institute of Physics for making possible the publication of the proceedings in this series. Finally, I thank Gerald T. Garvey, Director of the Argonne Physics Division for encouragement to proceed.

The Workshop was supported by the Division of Physical Research of the U.S. Energy Research and Development Administration.

> Gilbert J. Perlow
> Physics Division
> Argonne National Laboratory

vii

<u>CONTENTS</u>

viii

EXPERIMENTS IN THE ACCELERATOR BEAM: CHANGE
IN THE CHARGE RADIUS OF 2+ ROTATIONAL LEVELS

Stanley S. Hanna
Physics Department, Stanford University, Stanford, Ca. 94305 *

The method of in-beam implantation is discussed and illustrated by implantation of ^{57}Fe into single crystals of semiconductors. The application to isotopes which cannot be produced by radioactive sources is illustrated by a study of the isomer shifts in isotopic series of rotational nuclei. The microscopic theory of Meyer and Speth [1] gives dramatic agreement with the experiments especially in giving the negative values observed for some of these rotational levels.

The principal advantages of in-beam implantation Mössbauer experiments are the following: (1) The implantation process can be studied as a function of time, temperature, and other parameters, (2) very dilute sources can be prepared in any desired medium, and (3) the Mössbauer effect can be studied in any stable Mössbauer nucleus. If proton and deuteron (and perhaps alpha) beams are used, implantation directly into the target or its backing can be used since very little radiation damage is produced. However, these beams produce large backgrounds, which can be greatly reduced if Coulomb excitation by heavy ions is used. In this case the beam produces unacceptable radiation damage, and the method of implanting the nuclear recoils through vacuum into a medium not irradiated by the beam was developed at Stanford. [2] Early work with this method showed that in metals the implanted nuclei find substitutional sites and the Mössbauer spectra are usually identical to those obtained with conventional radioactive sources. In non-metals substitutional and "other" sites are produced, while in semiconductions two (or more) well-defined sites are observed. [3]

I would like to show an example of implantation into a semi-conductor. A typical arrangement in which the temperature can be varied is shown in Fig. 1 (low temperatures can be obtained by replacing the heating element by a liquid nitrogen or helium reservoir). Figure 2 shows spectra obtained for implantation of ^{57}Fe into single crystals of germanium as a function of temperature. Two well-defined sites are observed. The right hand resonance can be identified with a substitutional site, while the left hand resonance is produced by either an interstitial or a "damage" site. A complete account of this work can be found in the thesis of G. L. Latshaw. [3]

I would now like to turn to a series of experiments which illustrate the use of in-beam implantation to produce high-quality, single-line sources of nuclei which cannot be produced by radioactive parents. In particular, these experiments measure the isomer shifts in a complete series of isotopes. Usually only the proton-rich isotopes can be measured with radioactive sources; in-beam implantation

*Supported in part by the U. S. National Science Foundation.

can then be used to complete the series. Both the Gd and the Yb series
have been completed in this way.

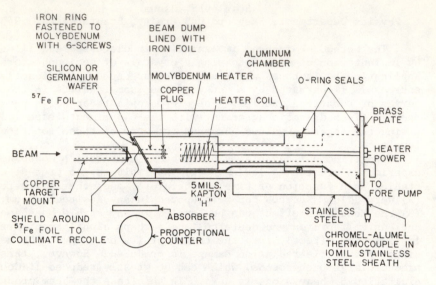

CROSS-SECTION DIAGRAM OF HEATER

Fig. 1.   Apparatus used for varying the temperature in
          in-beam implantation Mössbauer experiments.

Following are the important features of the Yb experiments:
  (1)   Only single-line sources and absorbers were used.
  (2)   Enriched targets and absorbers, while not essential, were
used for the rarer isotopes in order to conserve accelerator time.
  (3)   High purity (99.999%) Al was used as the implantation medium.
After testing Al, Pd, Ta, and Th, it was found that the cubic, non-
magnetic environment in Al gave the best combination of large f and
small Γ.
  (4)   To tie the implantation measurements on ^{172,174,176}Yb to
previous source experiments, ^{170}Yb was included in the series as a
source experiment, using neutron-irradiated TmAl$_2$ as the source.
These runs were carried out in a "conventional" Mössbauer cryostat.
  (5)   The implantation runs were made in an in-beam, nitrogenless
cryostat to expedite changing absorbers. Two target-catcher assemblies
were used in series, as shown in Fig. 3, to increase the source strength.
The complete assembly is shown in Fig. 4. The Mössbauer drive-rod in
both cryostats was under high tension to reduce compressional modes.
  (6)   The basic feature of the method was to compare absorbers A
and B simultaneously, as illustrated in Fig. 5. After a run in the
configuration AB the absorbers were interchanged and a run of equal
duration was made in the configuration BA. In this way each absorber
was exposed symmetrically to any possible imperfections in the
Mössbauer drive.
  (7)   A large number of absorbers were measured so as to be able

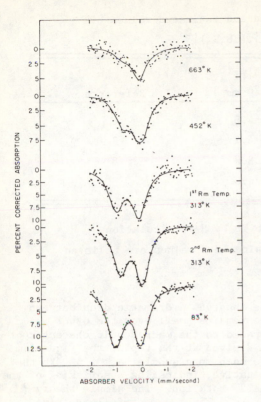

Fig. 2.   Mössbauer spectra of
^{57}Fe implanted into
germanium by means of a
64-MeV ^{35}Cl beam, at
various temperatures.

to select a suitable pair producing the maximum isomer shift. The results, shown in Fig. 6, indicate that Yb_2S_3 and YbF_2 produce the maximum shift and they also gave good symmetric single-line spectra.

(8) The absorbers used in the implantation runs were "calibrated" with the ^{170}Yb radioactive source. It was found that all YbF_2 absorbers gave identical ^{170}Yb shifts, while the Yb_2S_3 absorbers showed small but definite variations (probably due to imperfections in the absorbers). In the implantation runs these "empirical" shifts were then used in the analysis of the other isotopes (^{172,174,176}Yb), it being assumed that all Yb ions in a given absorber see the same average environment. In any case, these variations are not large enough to affect seriously the results.

Typical spectra are shown in Figs. 7 and 8. The major result of this experiment can be seen in these figures. In ^{170}Yb one can see a large (positive) shift between YbF_2 and Yb_2S_3, while in ^{176}Yb the shift is essentially zero. Thus, the addition of a few neutrons produces a profound effect on the size of the rotational 2^+ level relative to that of the ground state. In general the results from the AB configuration were slightly different from those in the BA configuration, indicating there might be a small effect from the Mössbauer drive. Averaging the AB and BA results should remove any such shift, but in any case the shifts were not large enough to influence seriously the results.

The final results are shown in Fig. 9 and compared with other experiments and with theory. At the top are plotted the ratios $\Delta<r^2>_A/\Delta<r^2>_{170}$ which need no calibration. At the bottom the values of $\Delta<r>_A$ are given which requires knowing the value of $|\psi(YbF_2)|^2 - |\psi(Yb_2S_3)|^2$. I will not give the details of arriving at a value for this quantity, but instead list in Table I some of the values that have been derived. Further details of the Yb results can be found in the thesis of P. B. Russell.[9]

In Fig. 10 the Yb results are shown together with results for the Gd series, two members of which were measured earlier with the

Table I. Various calculated and estimated values of
$$|\psi(\text{YbF}_2)|^2 - |\psi(\text{Yb}_2\text{S}_3)|^2$$

Method	$\Delta|\psi|^2$ 10^{-26} cm^{-3}	Reference
Hartree-Fock	4.8	6
0-ITS (Yb)	4.2	7
0-ITS (Sm)[a]	3.8	8
Adopted[b]	4.0	

[a] Value for Sm corrected to Yb by H-F calculations

[b] Average of 0-ITS values since H-F values are believed to be too high

implantation method.[10] In Gd one observes a dramatic drop between ^{154}Gd and ^{156}Gd and then the value remains small over the other isotopes. This behavior can be traced to the macroscopic character of the Gd rotors (note the change in β given at the top of the figure). In the Yb isotopes the changes are much more subtle; at ^{174,176}Yb the values probably become slightly negative, indicating a shrinking of the charge radius on excitation of the 2$^+$ rotation. The simple theory[5] fails to give either the smallness or the trend of the measured values. These properties are however dramatically explained by the microscopic theory of Meyer and Speth.[1] I would now like to give a brief outline of these theories.

A simple liquid drop model leads to a centrifugal stretching of the nucleus on excitation of the 2$^+$ rotation. Marshalek[5] applied the cranking model of Inglis with a

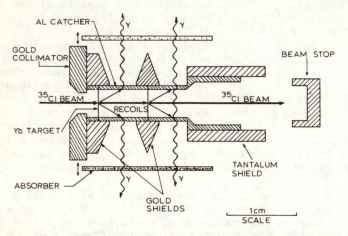

Fig. 3. Principal features of the method and the apparatus used for Coulomb excitation and recoil implantation through vacuum in a Mössbauer effect experiment. Note the series arrangement of two targets and two catchers that was used to increase the source strength.

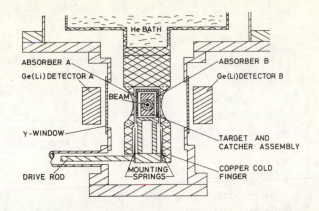

Fig. 4. Cross section drawing of the apparatus,
showing the velocity drive system, the
arrangement of targets and catcher, and
parts of the liquid helium Dewar. The
design of the apparatus made possible
the use of two simultaneous Mössbauer
detection systems. Heat radiation
shields and thermal grounding braids
have been omitted from the drawing for
clarity.

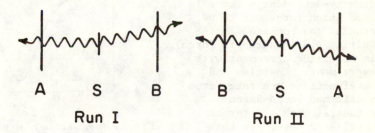

Fig. 5. Basic technique used in comparing isomer shifts
in Mössbauer absorbers. The first run is made
with absorbers in configuration AB followed by
a run in configuration BA, on either side of
the source S.

pairing plus quadrupole force. In this treatment one can identify
both a centrifugal and a coriolis term

$$\Delta \langle r^2 \rangle = \text{Cent} + \text{Cor}$$

but the former usually dominates the latter. Fair agreement with
experiment is obtained at the onset of deformation (cf. the Gd series
mentioned above) but the agreement is poor for the well-deformed

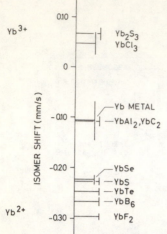

Fig. 6. A graphical representation of isomer shifts of the 84 keV γ-rays of ^{170}Yb for various absorber materials relative to a ^{170}TmAl$_2$ source. The vertical lines indicate the uncertainties in the measurements.

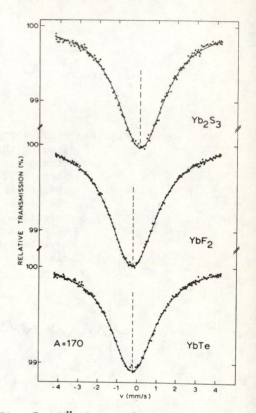

Fig. 7. Mössbauer absorption spectra of the 84 keV γ-rays of ^{170}Yb for the indicated ytterbium absorbers. The solid lines are least-squares fits of symmetrical Lorentzian lines to the data. The dashed lines indicate the centroids of the resonance dips.

nuclei (cf. the Yb series).

In the theory of Meyer and Speth,[1] the second-order cranking equations are developed within the framework of the Migdal theory of finite Fermi systems with the Migdal p-p and p-h interactions. The density-dependent δ force is used which turns out to be crucial (as compared to the pairing plus quadrupole force) in obtaining agreement with experiment. Energies and wavefunctions are taken from a deformed Woods-Saxon potential with a deformed oscillator base. In this theory superfluidity is achieved despite the strong nucleon interactions, and one can identify a coriolis antipairing (CAP) term in addition to the ones above

$$\Delta \langle r^2 \rangle = \text{Cent} + \text{Cor} + \text{CAP}.$$

It is this term which can be negative and large enough to dominate the others, so as to produce an actual shrinking of the nuclear charge radius. In the actual calculation no effective charge is used for the neutrons so that in effect the neutrons influence the charge radius only by their

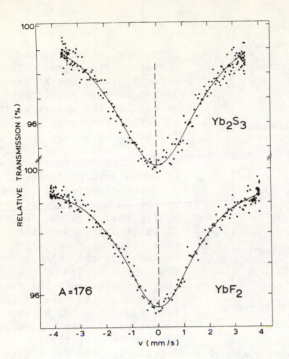

Fig. 8. Mössbauer absorption spectra of the 82 keV γ-rays of ^{176}Yb for absorbers of YbF$_2$ and Yb$_2$S$_3$. The solid lines are least-squares fits of dispersion modified Lorentzian lines ($2\xi = -0.03$) to the data. The dashed lines indicate the centroids of the resonance dips.

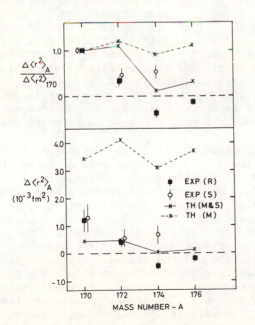

Fig. 9. Comparison of experimental and theoretical values of ratios of $\Delta\langle r^2\rangle$ and of $\Delta\langle r^2\rangle$ for the $0^+ \rightarrow 2^+$ transitions in the even-even ytterbium isotopes. The letters (R), (S), (M&S) and (M) signify the present results and those of Refs. 4, 1, and 5 respectively.

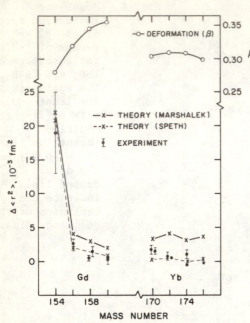

Fig. 10. Theories compared with
the measurements of
$\Delta\langle r^2\rangle$ for the isotopes of
Gd and Yb. The experi-
mental points are
identified in Fig. 9 and
Ref. 10.

"dragging" effect on the
protons. The overall theory
is equivalent to a Hartree-
Fock-Brueckner development
except for the nature of the
effective two-particle
interactions used.

The results and features
of the Meyer-Speth theory can
be summarized as follows:
(1) Polarization of the nucle
by the change in the electron
configuration is not important
in first order, any such
effects being inherently
included in the calculation.
(2) The rotation of the 2^+
level affects only the Nilsson
levels near the Fermi edge.
(3) Since the N+1 Nilsson
levels are larger than N level
but can be practically
degenerate, the CAP effect can
shift particles from N → N+1
levels or vice versa, thus
leading to either an expansion
or a contraction of the nuclea
size through the CAP term.
(4) In fact, the CAP effect
shifts particles from just
above to just below the Fermi edge, thereby enhancing the shrinking
effect. (5) In soft rotors (e.g., Gd) the centrifugal stretching
effect dominates, whereas in well-deformed nuclei (e.g., Yb) the CAP
effect is dominant, especially since the stretching effect is small in
these hard rotors. (6) Fine structure observed, as neutrons are
added to a series of isotopes, is a natural consequence of the details
of the Nilsson levels.

Predictions of the Meyer-Speth theory [1] for specific series of
isotopes are shown in Figs. 11-13. In the upper part of these figures
the ordinate gives the change in population of a given Nilsson orbit
on excitation of the 2^+ level. Open circles indicate orbits with N
greater than for solid circles. When the population of an N = 5 proto
orbit becomes larger at the expense of an N = 4 orbit the nuclear char
radius expands. On the other hand, we see that the reverse can occur
when N = 4 proton orbits increase in population at the expense of
N = 5 levels leading to a shrinking of the radius. The lower part of
these figures gives the change in charge radius as a function of A.

Figure 11 shows results for the Gd isotopes. Agreement with the
Mössbauer results in Fig. 10 is good. The dramatic decrease from
^{154}Gd to ^{156}Gd is reproduced, as well as the subsequent leveling off
for the heavier isotopes. The Dy isotopes are shown in Fig. 12. Larg
negative values are predicted but are not yet confirmed by experiment.

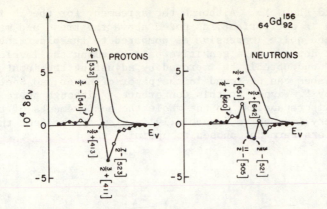

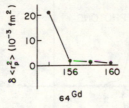

Fig. 11. Upper Part: The calculated change
of the occupation probabilities of
the Nilsson levels near the Fermi
energy for protons and neutrons in
^{156}Gd. The levels are drawn in a
schematic way, equally spaced in
the order of increasing energies.
Open circles refer to N = 5 protons
and N = 6 neutron levels, full
circles to N = 4 proton and N = 5
neutron levels, respectively. The
top lines give the occupation
probabilities in the ground state
(to different scale) showing the
diffuse Fermi edge.
Lower Part: The resulting $\Delta\langle r^2\rangle$'s
for the series of Gd isotopes.

Figure 13 shows Yb. Although the agreement for the Yb isotopes (see Fig. 10) is not perfect, as noted above the trend and magnitude are correct and quite impressive as compared to the older theory. In Yb the effect depends on a sensitive balance among the levels, and the fit could undoubtedly be improved by adjusting the input levels. Other comparisons can be found in the paper of Meyer and Speth [1] and in the talk by F. Wagner at this conference. Although there are some interesting discrepancies, on the whole the achievements of the theory are very impressive, especially in view of the fact that the input parameters were not chosen to optimize the fits.

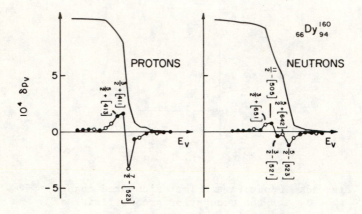

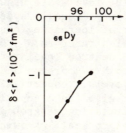

Fig. 12. Same as Fig. 11 except for the Dy isotopes.

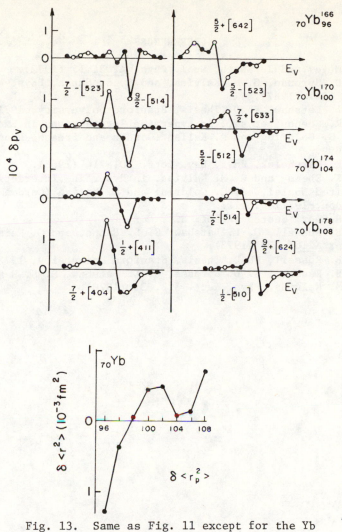

Fig. 13.   Same as Fig. 11 except for the Yb
           isotopes.

The experiments reported in this talk point the way for many
future investigations.   In the first place, it is important to measure
other rotational nuclei, especially those like Dy for which a large
negative shift is predicted.   Further measurements on nuclei like the
Yb isotopes, which are predicted to show fine structure, would be very
interesting.   Differences between the change in charge radii and mass
radii should be explored.   As mentioned by the authors, [1] there are
several refinements of the theory that could be made.   The process of
implantation can be studied in more detail.   In particular, measurements as a function of time should be made.   Controlled experiments
should be carried out to study the conditions which lead to the
occupation of substitutional sites as opposed to interstitial or
"damage" sites.

       In closing, I would like to thank my colleagues over the years,
P. B. Russell, G. Kaindl, G. L. Latshaw, G. D. Sprouse, G. M. Kalvius,
and B. B. Triplett, who have carried out these experiments.

REFERENCES

1. J. Meyer and J. Speth, Nucl. Phys. A203, 17 (1973).
2. G. D. Sprouse, G. M. Kalvius, and S. S. Hanna, Phys. Rev. Lett. 18, 1041 (1967).
3. G. L. Latshaw, Ph.D. Thesis, Stanford University, 1971, unpublished.
4. G. K. Shenoy, et al., in Hyperfine Interactions in Excited Nuclei, ed. G. Goldring and R. Kalish (Gordon and Breach, London, 1971), p. 699.
5. E. R. Marshalek, Phys. Rev. Lett. 20, 214 (1968).
6. G. K. Shenoy and G. M. Kalvius, in Hyperfine Interactions in Excited Nuclei, ed. G. Goldring and R. Kalish (Gordon and Breach, London, 1971) p. 1201.
7. S. Hüfner, quoted in Ref. 8.
8. P. B. Russell, G. L. Latshaw, S. S. Hanna, and G. Kaindl, Nucl. Phys. A210, 133 (1973).
9. P. B. Russell, Ph.D. Thesis, Stanford University, 1972, unpublished.
10. P. B. Russell, G. D. Sprouse, G. L. Latshaw, S. S. Hanna, and G. M. Kalvius, Phys. Lett. 32B, 35 (1970).

CHANGE OF THE MEAN SQUARE NUCLEAR CHARGE RADIUS FOR THE 2^+ - 0^+ TRANSITIONS IN ^{178}Hf, ^{182}W, and ^{186}Os

F.E. Wagner, M. Karger, M. Seiderer, and G. Wortmann
Physik Department, Technische Universität München,
D-8046 Garching, Germany

ABSTRACT

The change $\Delta\langle r^2\rangle$ of the mean square nuclear charge radius for the 2^+-0^+ rotational transitions in ^{178}Hf, ^{182}W, and ^{186}Os has been determined from Mössbauer isomer shift data for these isotopes in various cubic transition metal hosts. The results are $\Delta\langle r^2\rangle$ = +1 x 10^{-4} fm^2, -3.6 x 10^{-4} fm^2, and +2 x 10^{-4} fm^2 for the ^{178}Hf (93 keV), ^{182}W (100 keV), and ^{186}Os (137 keV) resonances, respectively.

INTRODUCTION

The Mössbauer isomer shifts that can be observed for the 2^+-0^+ Mössbauer transitions in deformed even-even nuclei in the tungsten region are, at best, a few percent of the natural linewidth. This is mainly due to the fact that the change $\Delta\langle r^2\rangle$ of the mean square nuclear charge radius is very small for low-energy rotational transitions. On the other hand, isomer shifts for such transitions are of interest for several reasons. First of all, the calculation of the $\Delta\langle r^2\rangle$ values for rotational transitions is a difficult task from the point of view of theoretical nuclear physics. Experimental $\Delta\langle r^2\rangle$ values thus provide a sensitive test for the microscopic model approach to this problem [1,2]. On the other hand, there are several elements, e.g. Yb, Hf, or W, for which only rotational transitions are available to the Mössbauer spectroscopist who wants to extract solid state information from the isomer shift. Although isomer shift measurements for such elements have little chance to ever find widespread application, some in-

formation on the solid state properties of favourable
systems can be obtained from careful experiments.

The instrumental difficulties arising in such measurements
can be overcome rather easily with modern Mössbauer spectrometers (see, for example, Ref. [3]). To obtain the data
reported here we have used a spectrometer in which three
absorption spectra are taken simultaneously in such a way
that zero-velocity drifts are reliably eliminated [4]. The
main problems one has to face in measurements of small
isomer shifts are thus (i) spurious center shifts as a consequence of unresolved hyperfine interactions in conjunction with texture in the samples or anisotropic f-factors,
(ii) the necessity to take into account the line asymmetry
arising from the dispersion term in the absorption cross-
section [3,5], and (iii) the second-order Doppler shift, which
may become of the same order of magnitude as the true
isomer shift.

The first of these problems can be overcome by the use of
cubic materials in order to eliminate hyperfine interactions as well as anisotropy effects arising from texture
or the f-factor. The second can be circumvented by comparing
the shifts in different sources as measured with identical
absorbers, because then the influence of the asymmetry of
the absorption lineshape cancels when the relative shifts
are evaluated. The remaining problem, namely the correction for the second order Doppler shift, cannot easily be
solved, since all practicable ways of determining the
second order Doppler shift from independent measurements
involve a model for the lattice dynamical properties of the
solid. This necessarily introduces a considerable degree
of uncertainty, and renders the influence of the second-
order Doppler shift the ultimate obstacle to accurate de-
terminations of small isomer shifts.

EXPERIMENTS AND RESULTS

In the experiments reported here we have carefully measured isomer shifts for the Mössbauer resonances in ^{178}Hf (93 keV), ^{182}W (100 keV), and ^{186}Os (137 keV) as sources in cubic transition metal hosts. The single-line absorber materials used in these experiments were HfC and HfV$_2$, W metal, and Os metal, respectively. All experiments were performed at 4.2 K.

In order to arrive at an estimate of the electron density differences in these hosts, we have made use of the systematics of the isomer shifts for transition elements as dilute impurities in transition metal hosts [6-8]. The isomer shifts for Mössbauer isotopes from the upper half of the 3d, 4d, and 5d transition series depend on the host in a very similar manner, as is illustrated by Fig. 1. The only Mössbauer isotope with a less than half-filled d shell in the metallic state for which large isomer shifts are observed is ^{181}Ta. Its isomer shifts do not, however, follow the pattern set by the heavier d transition elements (Fig.2). Still, it has a feature in common with all the other d Mössbauer isotopes, namely the fact that one always finds smaller electron densities in 5d hosts than in the 4d analogues [6-8]. This has previously been exploited for obtaining an electron density calibration for ^{181}Ta (6.2 keV) [7].

The results of our present isomer shift measurements for ^{182}W reveal a correlation with the ^{181}Ta data (Fig. 3). This suggests that the behaviour typical for the ^{181}Ta isomer shifts is common to the elements with less than half-filled d shells. For ^{178}Hf the isomer shifts in transition metal hosts are even smaller than for ^{182}W. Our data for this isotope thus do not unambiguously establish a correlation with the ^{181}Ta data, but at least they suggest that such a correlation exists (Fig. 4). To obtain

an estimate of the electron density differences at Hf and
W nuclei in transition metal hosts, we therefore assume

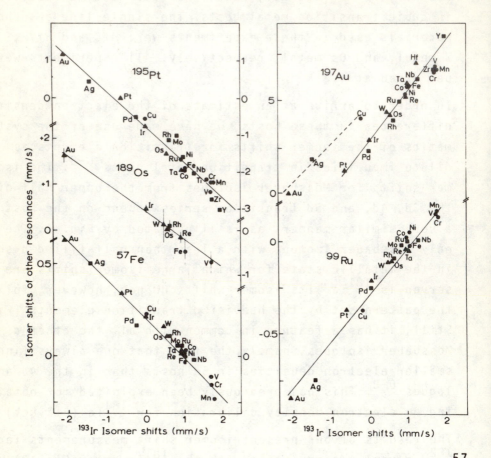

<u>Fig. 1</u>: Isomer shifts for the Mössbauer resonances in [57]Fe
(14 keV), [99]Ru (90 keV), [189]Os (36 keV), [195]Pt (99 keV) and
[197]Au (77 keV) as dilute impurities in transition metal
hosts, plotted versus the corresponding shifts for [193]Ir
(73 keV). Most of the data for [189]Os and [195]Pt have been
obtained in the course of the present work. The other data
are from Refs. [6,7] and the references given therein.

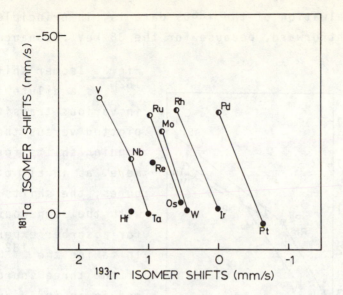

Fig. 2: Isomer shifts for ^{181}Ta as a dilute impurity in various transition metals, plotted versus the corresponding shifts for ^{193}Ir. The data points for homologous host metals from the 4d and 5d series are connected by straight lines in order to reveal the correlation described in the text.

that the ratios of the electron densities $\rho(0)_W / \rho(0)_{Ta}$ and $\rho(0)_{Ta} / \rho(0)_{Hf}$ have a reasonably well-defined value, and that this value can, at least approximately, be obtained from self-consistent calculations for free ions with the configurations $5d^n 6s^2$ and $5d^n$. Using the Dirac-Fock-Slater calculations of Kalvius and Shenoy [9] one thus finds values of about 1.2 for these ratios. With $\Delta \langle r^2 \rangle_{Ta} = -50 \times 10^{-3}$ fm^2 [7] and the average slopes of the lines connecting points for homologous hosts in Figs. 3 and 4, one then obtains $\Delta \langle r^2 \rangle_W = -3.6 \times 10^{-4}$ fm^2 and $\Delta \langle r^2 \rangle_{Hf} = +1 \times 10^{-4}$ fm^2.

The evaluation of the ^{186}Os data is, in principle, more straightforward, because for the 36 keV resonance in

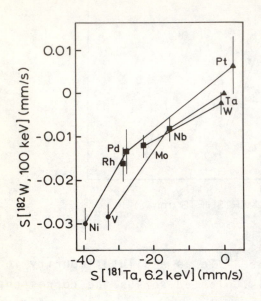

Fig.3: Isomer shifts for ^{182}W as a dilute impurity in various transition metal plotted versus the corresponding shifts for ^{181}Ta. Here, as in all other figures, the shifts are given with the sign appropriate for absorber experiments. In taking the ^{182}W data, on of the three sources compared in an individual run ha always been ^{182}Ta in Ta metal, which therefore serves as a common reference standard. Data points for homologous 3d, 4d, and 5d hosts have been connected by straight lines.

^{189}Os a value for $\Delta \langle r^2 \rangle$ has previously been determined[8,1] The ^{186}Os alloy isomer shifts (Fig. 4) could therefore directly be compared with the shifts for the 36 keV transi tion in ^{189}Os if ^{189}Os data for the same alloy systems ex- isted. Since this is not presently the case, we have used the ^{193}Ir isomer shift data as intermediaries between the two Os isotopes (Figs. 1 and 4). This procedure yields $\Delta \langle r^2 \rangle_{186Os} / \Delta \langle r^2 \rangle_{189Os} = -0.14 \pm 0.03$. With $\Delta \langle r^2 \rangle_{189Os} = -1.5 \times 10^{-3}$ fm^2 for the 36 keV transition in ^{189}Os as de- rived in Ref. [8] one then obtains $\Delta \langle r^2 \rangle_{186Os} = +2 \times 10^{-4}$ fm The data shown in Figs. 3 and 4 and used to derive the $\Delta \langle r^2 \rangle$ values, are all uncorrected for the second order Doppler shift. For ^{182}W we have performed f-factor measurements in order to obtain estimates of this contribution to the ex-

perimental shifts. The results are in fair agreement with
estimates from the Debye temperatures and show that the

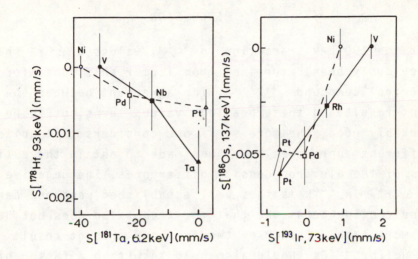

Fig.4: Isomer shifts for ^{178}Hf and ^{186}Os as dilute impurities in various transition metals plotted versus the corresponding shifts for ^{181}Ta and ^{193}Ir, respectively. The
data points connected by lines have been taken in separate
runs, each comparing a set of three different sources. The
data connected by fully drawn lines are presented for V as
the zero isomer shift standard, while for those connected
by dashed lines the standard is Ni. Therefore only data
from the same run can be compared directly.

corrections for the relative shifts in the studied alloys
are of the order of 0.005 mm/s. For ^{182}W this is still considerably smaller than the observed shifts and thus negli-
gible compared to the uncertainty inherent in the electron
density calibration. The ^{186}Os results are also virtually
unaffected by corrections of this magnitude. In the case of
^{178}Hf, however, the second order Doppler shift introduces
a considerable uncertainty, even though the correction with
our experimental values leaves the final result for $\Delta \langle r^2 \rangle$

virtually unchanged. Still, the uncertainties in the electron density calibration will be the dominant source of systematic error even in the case of ^{178}Hf.

DISCUSSION

For the 100 keV transition in ^{182}W, values of $\Delta \langle r^2 \rangle$ have previously been determined from isomer shift data for chemical compounds [8,11,12]. The agreement between the present result and these previous values turns out to be remarkably good, the more so if one considers that quite different approaches have been made to obtain the estimates of the electron density differences. The negative sign of $\Delta \langle r^2 \rangle$ in ^{182}W is thus well established, and one can with some confidence assume that the true value does not deviate by more than a factor of two from our present result. The value for ^{186}Os should also hold to within a factor of two, most of which is due to the large error that must be assigned to the $\Delta \langle r^2 \rangle$ value for ^{189}Os (36 keV). For ^{178}Hf larger uncertainties arise, mostly because of the assumptions made in the electron density calibration, but also because of the relatively large statistical errors and the influence of the second order Doppler shift. Still, the positive sign and the order of magnitude of $\Delta \langle r^2 \rangle$ can be considered as well established.

The ratios of $\Delta \langle r^2 \rangle$ for the first excited states in ^{184,186}W relative to ^{182}W have previously [8,11] been determined. They show that $\Delta \langle r^2 \rangle$ for these two isotopes is positive but somewhat smaller in magnitude than for ^{182}W. For ^{188,190,192}Os muonic isomer shift data have yielded increasingly negative $\Delta \langle r^2 \rangle$ values [13,14]. Thus a considerable amount of data is now available for a comparison with the calculations of Meyer and Speth [1,2]. It turns out that the theory correctly reproduces the extremely small order of magnitude of $\Delta \langle r^2 \rangle$ for the Hf and W isotopes, but that it does not give the exact values, or even the right signs, with any confidence.

The calculations of Meyer and Speth unfortunately do not cover ^{186}Os. They do, however, correctly reproduce the increasingly negative muonic isomer shifts for ^{188,190,192}Os. For ^{188}Os (155 keV) and ^{190}Os (187 keV) it should be possible to measure $\Delta \langle r^2 \rangle$ by Mössbauer spectroscopy. This would be of interest, particularly since the accuracy to which muonic isomer shifts are a measure of $\Delta \langle r^2 \rangle$ has been a matter of discussion [13,14].

ACKNOWLEDGEMENT

This work has been supported in part by the Federal Ministry for Research and Technology of the Federal Republic of Germany. The ^{178}W sources used in the ^{178}Hf experiments were produced at the Cyclotron Laboratory of the Gesellschaft für Kernforschung, Karlsruhe. This support is also gratefully acknowledged.

REFERENCES

1.   J. Meyer and J. Speth, Nucl.Phys. A 203, 17 (1973)

2.   J. Speth, W. Henning, P. Kienle and J. Meyer, in "Mössbauer Isomer Shifts", ch. 13 (eds. G.K. Shenoy and F.E. Wagner, North Holland Publishing Co., Amsterdam, 1977

3.   G.K. Shenoy, F.E. Wagner and G.M. Kalvius, ibid. ch.4

4.   G. Kaindl, M. R. Mayer, H. Schaller and F.E. Wagner, Nucl. Instr. Methods 66, 277 (1968)

5.   W. Potzel, F.E. Wagner, G.M. Kalvius, L. Asch, J.C. Spirlet and W. Müller, contribution to this conference, and references given therein

22

6. F.E. Wagner, G. Wortmann and G.M. Kalvius, Phys. Letters 42A, 483 (1973)

7. G. Kaindl, D. Salomon and G. Wortmann, Phys.Rev. B 8, 1912 (1973)

8. F.E. Wagner and U. Wagner, in "Mössbauer Isomer Shifts", ch. 8a (eds. G.K. Shenoy and F.E. Wagner), North Holland Publishing Co., Amsterdam 1977

9. G.M. Kalvius and G.K. Shenoy, Atomic Data and Nuclear Data Tables 14, 639 (1974)

10. F.E. Wagner, D. Kucheida, U. Zahn and G. Kaindl, Z. Physik 266, 223 (1974)

11. F.E. Wagner, H. Schaller, R. Felscher, G. Kaindl and P. Kienle, in "Hyperfine Interactions in Excited Nuclei", Vol. 2, p. 603 (eds. G. Goldring and R. Kalish) Gordon and Breach, New York 1971

12. H. Bokemeyer, K. Wohlfahrt, E. Kankeleit and D. Eckardt, Z. Physik A 274, 305 (1975)

13. H. Backe, E. Kankeleit and H.K. Walter, in "Mössbauer Isomer Shifts" ch. 14 (eds. G.K. Shenoy and F.E. Wagner) North Holland Publishing Co., Amsterdam 1977

14. H.K. Walter, Nucl.Phys. A 234, 504 (1974)

NEW RESULTS WITH ^{73}Ge: DISPERSIVE INTERFERENCE EFFECTS AND THE 13.3 keV MÖSSBAUER RESONANCE AT NATURAL LINEWIDTH

Loren Pfeiffer
Bell Laboratories
Murray Hill, New Jersey 07974

ABSTRACT

Lithium-drifted silicon and intrinsic germanium detectors are compared as detectors for the recoilless γ-rays from ^{73}Ge at 13.3 keV. Use of germanium detector results in improved background rejection and allows one to observe large Mössbauer resonance effects. The unbroadened Mössbauer resonance from the 13.3 keV level of ^{73}Ge is observed. The resonances all show a dispersive lineshape asymmetry due to final state interference effects between photoelectric absorption and Mössbauer absorption followed by internal conversion.

INTRODUCTION

The introduction of liquid nitrogen cooled intrinsic germanium (IG) counters has recently made it possible to detect the ^{73}Ge 13.3 keV γ-rays with a much reduced X-ray background. This is illustrated in Fig. 1 using the ^{73}Ge parent source, ^{73}As. Spectrum 1a was obtained with a lithium compensated silicon (Si(Li)) detector, and 1b is the same radiation as seen by the IG detector. As is seen, comparing 1a and 1b, both detectors have similar energy resolution, but the background in the vicinity of the 13.3 keV Mössbauer γ-ray is $\sim \times 4$ smaller in the IG spectrum.

As discussed elsewhere[1] the lower background in the IG detector is due to a reduced rate for Compton events from the 54 keV γ-rays which accompany the γ-rays of interest. This reduced Compton rate in the IG detector is due to the competition resulting from the larger cross-section for events in the 54 keV photopeak.

This improvement and others have made it possible to observe ^{73}Ge recoilless resonance effects as large as 6.0(2%) with FWHM linewidth 13.9(5) μm/sec, using germanium single crystal absorbers enriched in ^{73}Ge. Representative of current ^{73}Ge Mössbauer spectroscopy is the spectrum in Fig. 2.

This spectrum was obtained using a special absorber 1 cm^2 by 25 μm thick made of single crystal Ge enriched to 79% in ^{73}Ge. The absorber crystal was epitaxially grown[2] by vacuum deposition on a (111) single crystal substrate of natural Ge. The substrate which was held at 890°C during the vacuum deposition was later completely removed by polishing and chemical etching leaving only the enriched ^{73}Ge (111) single crystal film.

24

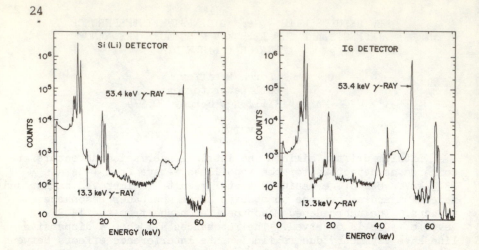

Fig. 1. Comparative γ-ray spectra using IG and Si(Li) detectors in a ^{73}Ge Mössbauer experiment. Part (a) is the γ-ray spectrum taken with a Si(Li) detector of 80 mm^2 × 5 mm. Part (b) is the same radiation as seen using an IG detector of 100 mm^2 × 5 mm. For both spectra the γ-ray geometry and the electronics were invariant, only the detectors and their preamplifiers were interchanged. Both spectra were obtained with a source (1 mCi ^{73}As in a Ge single crystal matrix) and absorber (enriched ^{73}Ge single crystal) in standard γ-ray transmission Mössbauer geometry.

A least squares computer fit of the resonance in Fig. 2 reveals a definite asymmetry in the resonance lineshape. Following the formalism worked out for the well-known lineshape asymmetry in ^{181}Ta, the data were fit with a dispersive lineshape,[3,4]

$$N(v) = N(\infty)(1-A(1-2\xi x)/(1+x^2)), \qquad (1)$$

where $x = 2(v-S)/W$. Here $N(v)$ is the γ-ray intensity transmitted through the absorber with the source moving at Doppler velocity v, S is the isomer shift, W is the linewidth, A is the resonance effect, and ξ is a measure of the relative size of the dispersion term.

In ^{181}Ta the dispersion is known to be the result of an interference between photoelectric absorption and Mössbauer absorption followed by internal conversion. In the case of ^{73}Ge the 13.3 keV transition is E2 and the atomic photoeffect is predominantly E1 so one might expect that any interference between them might be small. As developed more completely in Ref. 5, however, Hannon[6] has shown that ξ goes as $(\alpha/10\pi\lambdabar^2)^{1/2}$ for an E2 transition. Using this we have pointed out[5] that the product $\alpha\lambdabar^{-2}$ is 110 times larger for ^{73}Ge than for ^{181}Ta which makes it reasonable to expect a comparatively large dispersion effect for ^{73}Ge.

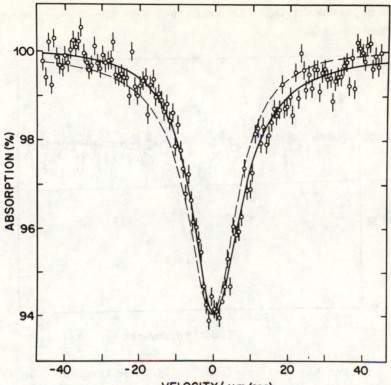

Fig. 2. Mössbauer effect from the 13.3 keV level of [13]Ge using
an isotopically enriched absorber. Spectrum taken in γ-ray
transmission geometry with an intrinsic germanium γ-ray detector,
pulse pileup rejection electronics, and source made by diffusing
[73]As into a natural (111) Ge crystal, (see Ref. 2). The
resonance absorber is a (111) Ge single crystal 25 μm thick by
1 cm², grown by vacuum deposition of Ge enriched to 79% in [73]Ge.
The solid line is the least squares fit to the data using Eq. (1).
The dashed line is not a fit but a comparison curve which
illustrates the lineshape asymmetry of the data. It is obtained
by reversing the sign of the dispersion term from the solid curve.
The dispersion effect may be seen by noting that the data points
are well fit by the solid line and not fit at all by the dashed
line.

The reduced γ-ray background has also made feasible experiments with unenriched germanium (7.76% [73]Ge) absorbers. An experiment using a high quality zone refined, but nonenriched Ge crystal
as Mössbauer absorber gives a 1.2(1)% effect at 6.9(9) μm/sec FWHM
linewidth, (see Fig. 3). This linewidth is equal within statistics
to the uncertainty principle limited width of $2\Gamma_0 = 2\hbar c/\tau E_\gamma =$
6.98(6) μm/sec. The natural linewidth resonance also shows a

lineshape asymmetry which is consistent with the value 2ξ quoted above.

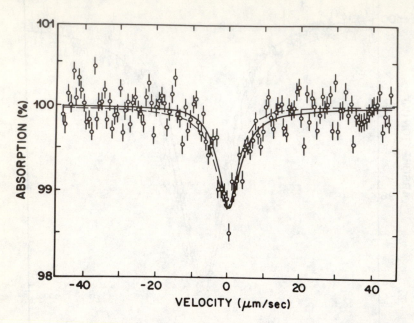

Fig. 3. Mössbauer effect from the 13.3 keV level of ^{73}Ge using a zone refined Ge crystal absorber made from isotopically natural material, (^{73}Ge isotopic abundance 7.76%). Except for the resonance absorber all conditions were exactly as in Fig. 2. The natural absorber used here was also a (111) Ge crystal 25 μm thick by 1 cm^2. The solid and dashed lines have the same meaning as in Fig. 1.

REFERENCES

1. Loren Pfeiffer, Nucl. Instr. Meth. 140, 57 (1977).
2. Loren Pfeiffer, R. S. Raghavan, C. P. Lichtenwalner, and A. G. Cullis, Phys. Rev. B12, 4793 (1975).
3. G. T. Trammell and J. P. Hannon, Phys. Rev. 180, 337 (1969).
4. G. Kaindl, D. Solomon, and G. Wortmann, Phys. Rev. B8, 1912 (1973).
5. Loren Pfeiffer, Phys. Rev. Lett. 38, 862 (1977).
6. J. Hannon, Private communication. See also Ref. 5.
7. H. C. Goldwire, Jr., and J. P. Hannon, (to be published).

THE QUADRUPOLE INTERACTION IN ZINC METAL

W. T. Vetterling and R. V. Pound
Harvard University, Cambridge, Ma. 02138

ABSTRACT

The 93.3 keV level of ^{67}Zn, with 9.1 μsec half-life, has a natural fractional linewidth of 10^{-15}, and provides the narrowest Mossbauer resonance yet observed. Our past investigations of ^{67}Zn have been based on the observation of resonances using an enriched ZnO absorber, with sources made by diffusing ^{67}Ga into sintered pills of zinc chalcogenides. Sintered copper hosts have also been used, and natural sphalerite has been employed, in one instance, as an absorber.[1]

To allow measurement of the quadrupole interaction in zinc metal, the enriched ZnO was reduced to zinc metal powder and compressed into a pill of thickness 1.4 gm/cm^2. Sources were made by diffusing 20 mCi of ^{67}Ga into sintered copper pills as described in our previous report.[1] The apparatus was also that described previously; in particular, the transducer was based on a cylinder of PZT-4 with ½-inch length, and could cover linearly a velocity range of ± 100 μ/s at 200 Hz. The multiscalar was a modified Northern model NS600, with a minimum dwell time of 20 μs, and with a 10-count buffer at the input to eliminate deadtime from memory cycling (for count rates of up to about 3×10^5 counts/s).

An early run with a velocity scan of ± 129 μ/s showed a resonance with hyperfine splitting at approximately -112 μ/s, which is presumed to be ZnO. (Occlusions of ZnO were, in fact, subsequently observed with a microscope.) Evidence also existed in this run of several lines at lower source velocities. A second run with a range of ± 70 μm/s showed a three-line spectrum with the 2:1 spacing expected for the spin-5/2 ground state of ^{67}Zn. Lines were observed at source velocities of -29.1 ± 0.9 μ/s, +24.8 ± 0.8 μ/s, and +51.8 ± 1.7 μ/s. The isomer shift was 15.8 ± 0.6 μ/s and the quadrupole interaction was $e^2qQ/h = 13.5 \pm 0.4$ MHz. This run performed with a transducer drive frequency of 100 Hz required a transducer voltage of 125 volts rms thirty percent beyond the linear range of the device. As a result, the line at highest velocity was somewhat broadened.

A third run with a source velocity range of ± 41.9 μ/s was within the transducer's linear range, and the results showed two of the three lines at velocities of 25.8 ± 0.8 μ/s and -29.3 ± 0.9 μ/s. A total count of 346×10^6 per channel was collected with 256 channels of data, and the line characteristics are shown in Table I.

Table I Line Parameters

Resonant Depth = 0.04%
Isomer Shift = 16.6 ± 0.5 μ/s
Linewidth = 2.4 μ/s
$e^2qQ/h = 13.8 \pm 0.4$ MHz

The isomer shift is quoted relative to a zinc absorber, and is much smaller in magnitude than that which would be measured relative to the more commonly used ZnO absorber. The result for the quadrupole interaction can be combined with previous measurements[2-4] to arrive at the results:

$$\frac{q_{Zn}}{q_{ZnO}} = 5.7 \pm 0.2 \qquad \frac{Q_{9/2}}{Q_{5/2}} = 3.6 \pm 0.2$$

where q_{Zn} and q_{ZnO} are the field gradients in zinc metal and zinc oxide respectively, $Q_{5/2}$ is the quadrupole moment of the spin-5/2 ground state, and $Q_{9/2}$ is the quadrupole moment of the spin-9/2 metastable state at 602 keV. Using the adopted value[5] $Q_{5/2} = 0.17$ b, we can also conclude that $Q_{9/2} = 0.61$ b, $q_{Zn} = 2.3 \times 10^{24}$ cm^{-3}, and $q_{ZnO} = 4.1 \times 10^{23}$ cm^{-3}. In this final run, the lines were observed to have a frequency separation of 4.14 ± 0.12 MHz. An attempt was, therefore, made to observe the nuclear quadrupole resonance. A signal-to-noise calculation using a comparison to the observed resonance of hexamethylene-tetramine at 3.31 Mhz showed that the signal-to-noise with a natural sample of ^{67}Zn was likely to be less than one at room temperature. Indeed, an investigation of 325-mesh zinc powder, and of zinc dust, from 3.4-4.2 MHz produced no lines. Experiments at 4.2^{O}K were hampered by strong RF-absorption, which required that the sample be diluted to fill only 1/6 of the available RF coil volume. Both natural zinc (325-mesh) and enriched zinc (produced by powdering and annealing the Mossbauer absorber) were scanned from 3.6-4.5 Mhz with no success. It is suspected that the ZnO contamination in the enriched Zn may cause serious inhomogeneity of the quadrupole interaction. A study of the theory of RF absorption by spherical metallic particles[7] suggests that a sample of well-annealed particles of less than 1 μm diameter may still hold promise for success in finding the zero-field quadrupole resonance.

REFERENCES

1. D. Griesinger, R. V. Pound, and W. T. Vetterling, Phys. Rev. B15, 3291 (1977).
2. H. Bertschat, E. Recknagel, and B. Spellmeyer, Phys. Rev. Lett. 32, 18 (1974).
3. G. J. Perlow, W. Potzel, R. M. Kash, and H. deWaard, J. Phys. (Paris) 35, C6-197 (1974).
4. C. Hayes, Thesis, Harvard University (1973).
5. G. Fuller and V. W. Cohen, Nuclear Data A5, 433 (1969).
6. G. D. Watkins and R. V. Pound, Phys. Rev. 85, 1062 (1952).
7. E.L. Sloan, Thesis, Harvard University (1962).

Piezoelectric Mössbauer Spectrometer With Fast Channel Advance Rates[*]

A.A.Forster, W.Potzel, G.M.Kalvius

Physik- Department, Technische Universität München

D- 8046 Garching, Germany

ABSTRACT

A Mössbauer spectrometer is described which uses a
piezoelectric quartz drive to generate Doppler velocities
for high resolution Mössbauer experiments with the 93.3 keV
resonance in ^{67}Zn. The spectrometer is coupled to a PDP8e
computer by an interface which initiates a read in cycle
when a nuclear pulse has been counted in the detector. The
interface works free of dead time at mean countrates $\leq 2 \cdot 10^5 s^{-1}$
and allows channel advance frequencies up to some 10 MHz.

INTRODUCTION

In Mössbauer measurements using the 93.3 keV γ-transition
in ^{67}Zn with its extraordinarily small fractional width
($\Gamma_o/E \approx 5 \cdot 10^{-16}$), piezoelectric transducers have been employed
most successfully. The task is to generate Doppler velocities
in the range from ≈ 1 μm/s (minimum observable width is
0.31 μm/s) to a few 100 μm/s (because of large hyperfine
interactions). For this purpose drive motors have been
developed which make use of the piezoelectric effect in
quartz[1,2,3] or in lead zirconate-titanate based materials[4].
By applying a sinusoidally varying voltage much below the resonance frequency of the piezoelectric crystal a velocity
sweep is generated. To record the spectra automatically the
Doppler velocity sweep is then synchronized with the channel

* Supported in part by the Bundesministerium für Forschung
  und Technologie

advance sweep of a multichannel analyzer like in standard Mössbauer spectroscopy where electromechanical or mechanical drive systems have been used. In particular the large isomer shifts which have been observed in various zinc chalcogenides [4,5], demand that in the investigations of unknown materials velocity ranges in excess of 100 μm/s will have to be covered. To obtain Doppler velocities of that magnitude the piezo-electric drive has to be operated at frequencies of several hundred Hz since the maximum obtainable amplitude is limited by the voltage which can safely be fed to the transducer and by the mechanical stress which the piezoelectric crystal can take continuously without rupture. It should be kept in mind that the transducer is located inside a He cryostat and operated at 4.2 K or below. The relatively high frequency puts a serious limit on the number of channels to be used to store the Mössbauer spectrum, since the minimum channel dwell-time is ≥ 3 μs for most multiscaler systems. This time span is governed by the memory cycle times and does not improve sub-stantially even if solid state memories are used.

In this communication we describe a spectrometer where spectra can be recorded with channel dwell-times well below 1 μs by making use of a different logic to transfer the counts recorded into the proper address location.

EXPERIMENTAL

The quartz drive used is based on a design described by Perlow [6]. Some modifications were carried out in particular with respect to the position of the three quartz crystals in order to reduce unwanted lateral motion to a minimum. Velocity calibration was performed with a source of (^{67}Ga) ZnO and a ZnO absorber enriched to 85.2% in ^{67}Zn by comparing the observed hyperfine splitting with the known [7] quadrupole interaction in ZnO.

The data pulses were collected in 512 channels of the memory of a PDP8e computer via the interface described in

principle in [8]. Additional modifications were incorporated
to the interface in order to make possible channel dwell-
times ≤ 1μs without significant losses of counts. A block
diagram of the modified part of the interface is shown in
Fig. 1. It replaces the unit shown in Fig. 3 of ref. 8.

OPERATION OF SYSTEM

The channel advance pulses (CA) drive a 12 bit scaler
(address counter) in phase with the drive voltage reference
signal. Exact phasing is ensured by the start (ST) pulse via
a synchronization logic, which sets the address counter to
zero at the beginning of each velocity sweep. When a nuclear
count arrives at the DET input of the read in control logic,
the momentary address number of the address scaler is trans-
ferred to a 64x12 bit fifo register [9]. This sequence will not
interrupt the advance of the address scaler. Subsequently a
read out cycle of the fifo register is initiated by the read
out control logic and the content of the channel corresponding
to the address number stored in the fifo register is increased
by one. This last step is at present the slowest sequence in
the data storage process. Since the computer memory operates
in double precission the write in cycle needs slightly less
than 4 μs. Clearly the mean rate of nuclear pulses arriving
at the DET input must not exceed $2.5 \cdot 10^5$ s^{-1}. On the other
hand, this rate can fully be used without loss of data. The
read in - read out cycles of the fifo register work completely
independent of each other with a frequency limit of 10 MHz.
Thus, if new detector pulses arrive during the time span of
the memory write in cycle, the corresponding addresses are
simply stored in the fifo register. Its storage capacity is
64 addresses of 12 bit wordlength which exceeds all possible
demands. Therefore statistical fluctuations in the mean
separation of 4 μs of the detector pulses at the maximum
allowable countrate are of no concern. The channel advance
frequency is limited only by the speed of the address counter
which at present is a few 10 MHz. It should be noted that the
250 kHz limit on the mean detector pulse rate are no obstacle

in our present system of recording ^{67}Zn spectra. Since we use
a NaJ(Tl) detector with energy selection, the 250 kHz limit
is higher than the maximum possible countrate in the 93 keV
window. It is, however, possible to increase this rate by
employing a fast intermediate storage memory between the
interface and the computer memory.

The channel selector shown in Fig. 1 provides the
possibility to choose the number of channels for the storage
of the spectrum between 128 and 4096.

RESULTS

Fig. 2a shows a Mössbauer absorption spectrum obtained
with a source of ^{67}Ga in a ZnO single crystal [3] and an
absorber of polycrystalline ZnO, enriched to 85.2% in ^{67}Zn.
The intensity ratios of the 5 lines are in good agreement
with theory [10]. The spectrum displayed in Fig 2b was recorded
with a source of ^{67}Ga in ZnS and an absorber of ZnS, enriched
to 87.9% in ^{67}Zn. The ^{67}Ga ($T_{1/2}$=78h) activity was obtained
in situ by 11MeV deuteron bombardment of a cubic ZnS single
crystal, which afterwards was annealed in a H_2S atmosphere
at elevated temperatures. The ZnS powder absorber was annealed
at about 1400 K also in a H_2S atmosphere. The relatively
broad absorption dip in Fig. 2b is probably due to heavy
radiation damage in the source. Also a partial transformation
of ZnS from the cubic into the hexagonal modification during
the cyclotron bombardment cannot be ruled out.

Acknowledgment:

We wish to thank Dr. E.Huenges,R. Warmt and the cyclotron
group in Munich for performing the irradiation and Dr.U.Wagner
for preparing the ZnS absorber.

REFERENCES

1.   H.de Waard and G.J.Perlow,
     Phys. Rev. Lett. $\underline{24}$, 566 (1970)

2.   A.I.Beskrovny, N.A.Lebedev, and Y.M.Ostanevich,
     Joint Inst. for Nuclear Research report, Dubna,
     1971 (unpublished)

3.   W.Potzel,A.Forster, and G.M.Kalvius,
     Journ. Physique Colloq. $\underline{C6}$, C6-691 (1976)

4.   D.Griesinger, R.V.Pound, and W.Vetterling,
     Phys. Rev. $\underline{B15}$, 3291 (1977)

5.   A.Forster, Diplomarbeit, Technische Universität
     München, 1976

6.   G.J.Perlow, in Perspectives in Mössbauer Spectroscopy,
     S.G.Cohen and M.Pasternak, eds.
     (Plenum Press, New York), 1972, pp. 221-237

7.   G.J.Perlow, W.Potzel, R.M.Kash, and H.de Waard,
     Journ. Physique Colloq. $\underline{C6}$, C6-197 (1974)

8.   A.Forster, N.Nalder, G.M.Kalvius,W.Potzel, and L.Asch
     Journ. Physique Colloq. $\underline{C6}$, C6-725 (1976)

9.   Data Sheet, Monolithic Memories, Device 57401/67401
     (1976)

10. G.J.Perlow, L.E.Campbell, L.E.Conroy, and W.Potzel,
     Phys. Rev. $\underline{B7}$, 4044 (1973)

34

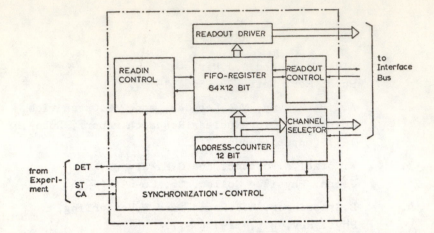

Fig.1 Modification of the PDP8e interface
for channel dwell-times ≤1 μs

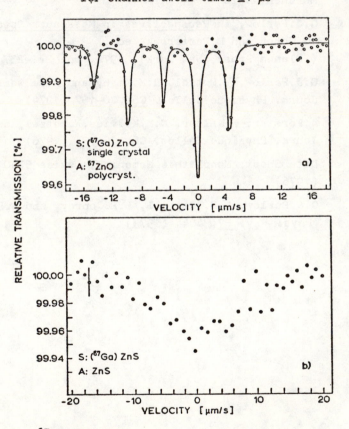

Fig.2 ^{67}Zn Mössbauer spectra

a) single crystal ZnO source and enriched
polycrystalline ZnO absorber

b) ZnS source and enriched ZnS absorber
(The data has been smoothed by averaging
adjacent channels)

MÖSSBAUEREFFECT WITH ^{137}La

E. Gerdau, H. Winkler, F. Sabathil

II. Institut für Experimentalphysik,

University of Hamburg

ABSTRACT

Mössbauereffect with a radioactive absorber of $6 \cdot 10^4$y ^{137}La was observed. Spectra for the system CeO_2-$LaCrO_3$ and CeO_2-La_2O_3 have been fitted and yielded values for the quadrupole interactions and the ratio of the quadrupole moments of the ground and excited state.

INTRODUCTION

^{137}La (fig.1) is an attractive nucleus for the observation of the Mössbauer Effect as the first excited state at 10,1 keV has a half life of 89ns. Thus high resolution measurements are possible where the nonmagnetic Lanthanum serves for example as a probe in various magnetic rare earth compounds.

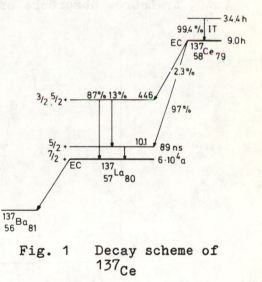

Fig. 1   Decay scheme of ^{137}Ce

EXPERIMENTAL

About 4 mg of $6 \cdot 10^4$y ^{137}La were produced by neutron activation of 1 kg of natural Cerium and then chemically separated. Finally absorbers of La_2O_3 and $LaCrO_3$ were prepared.

Our enriched CeO_2 sources contained 21,7% ^{136}Ce. With sources of natural Ce the 10,1 keV could not be observed.

All measurements were performed at room temperature. A conventional sinusoidial Mössbauer drive was used which operated at the resonance frequency.

RESULTS

The results of two measurements with CeO_2-sources and La_2O_3 or $LaCrO_3$ absorbers are displayed in fig.2 and 3.

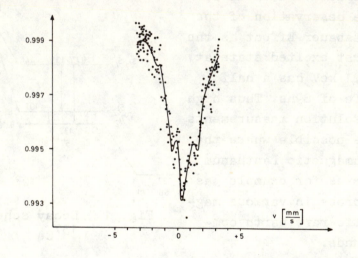

Fig. 2   Mössbauer spectrum with CeO_2
          source and $LaCrO_3$ absorber

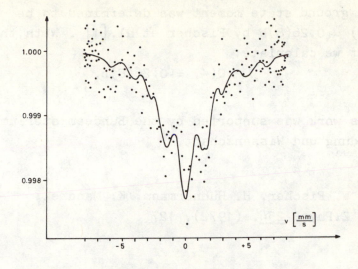

Fig. 3   Mössbauer spectrum with CeO$_2$
source and La$_2$O$_3$ absorber

A transmission integral fit was possible in both cases.
The parameters for the fit were the normalization
constant, the effect, the isomeric shift, the quadrupole interaction in the excited state in source and absorber and the ratio of the quadrupole moments. The
results are given in table I.

Table I   Results of least square fits

system	eQ(source)V$_{zz}$ [10^{-6} eV]	eQ(abs)V$_{zz}$ [10^{-6} eV]	$\frac{Q(\frac{5}{2})}{Q(\frac{7}{2})}$
CeO$_2$-La$_2$CrO$_3$	+ 0.632(28)	+ 0.400(36)	1.005(30)
CeO$_2$-La$_2$O$_3$	+ 0.620(32)	+ 0.736(24)	0.948(30)

38
The ground state moment was determined to be
$Q(\frac{7}{2})$ =+0.26(8)b by Fischer et al.[1] . With this result we calculate:

$$Q(\frac{5}{2}) = +0.26(8)b.$$

This work was supported by the Bundesmisterium für Bildung und Wissenschaft.

[1] W. Fischer, H. Hühnermann, K. Mandrek;
Z.Phys. 254, (1972), 127

SOME RECENT DEVELOPMENTS IN THE
^{181}Ta(6.2 keV) - SPECTROSCOPY

G. Wortmann[*], V. Dornow, A. Heidemann, G. Trollmann
Physik-Department, Technische Universität München
D-8046 Garching, W. Germany

ABSTRACT

The 6.2 keV level of ^{181}Ta($T\frac{1}{2}$ = 6.8 μs) belongs to the few Mössbauer resonances with lifetimes in the microsecond region. The corresponding natural linewidth, W_o = 6.5 μm/s, has not been observed so far. In spite of many efforts, the best experimental linewidths are still a factor of ~10 larger than W_o[1,2]. In this contribution we report a new method of source preparation from cyclotron-produced, carrier-free ^{181}W activity, which yields either an improvement in the experimental linewidth or in the source efficiency. The hitherto method of source production from neutron activated ^{181}W(^{180}W(n,γ) ^{181}W) suffered from a relatively low specific activity, where even with high-flux neutron irradiations only ~100 mCi/mg could be obtained[3]. In contrast, a ^{181}W production by a deuteron (or proton) bombardment of a Ta foil (^{181}Ta(d,2n) ^{181}W) followed by a chemical separation of the carrier material (by ion exchange chromatography) can provide specific activities up to ~4 000 mCi/mg. From 15 mCi of carrier-free ^{181}W activity, several sources were produced by diffusing the activity into single crystals of W and Mo[4]. With a ^{181}W($\underline{W}$) source and a 7.2 mg/cm^2 thick Ta metal absorber we observed a resonance line, which exhibited the best experimental linewidth observed so far, W = 54(1) μm/s, when fitted with a single Lorentzian modified by a dispersion term with a (fixed) value of 2ξ = 0.35. Due to the higher specific activity the mean diffusion depth of the ^{181}W activity in the source matrix could be reduced. This lead to a considerable improvement in the source efficiency because of the smaller photoabsorption of the 6.2 keV gamma rays. With a ^{181}W($\underline{Mo}$) source, for example, we observed a resonance effect twice as large as with sources from neutron-activated ^{181}W activity[3].

Due to the extreme sensitivity of the ^{181}Ta resonance to isomer shifts, most of the resonance spectra show large lineshift-to-linewidth ratios, which can amount up to ~100 [3]. This leads to an enormous loss in effective measuring time, when standard velocity spectrometers are used, which sweep the whole velocity range. We present some spectra obtained with a region-of-interest spectrometer which scans only the velocity range around the resonance and which overcomes thereby the difficulties just mentioned. The velocity resolution of our spectrometer was better than $2 \cdot 10^{-3}$. The best application of such a region-of-interest spectrometer is the study of temperature or pressure effects on the hyperfine parameters. We studied the temperature dependence of the isomer shifts S in $KTaO_3$ and $NaTaO_3$ between 4.2 K and 700 K. Between 200 K and 700 K we found a linear variation of S with a slope of $(\delta S / \delta T)_p$ = 48 (5) μm/s· K and 45(5) μm/sK, respectively. These figures are 20 times larger and of the opposite sign than the second-order Doppler shift and demonstrate strikingly the extreme sensitivity of the ^{181}Ta resonance to changes of the electron density. An analysis of the data in terms of an implicit temperature dependence (due to the volume expansion) and an explicit temperature dependence (which was observed in metall system with the ^{181}Ta resonance [5]) must await for data on the pressu dependence of S.

ACKNOWLEDGEMENTS

This work was supported by the Bundesministerium für Forschung und Technologie.

REFERENCES

1. G. Wortmann, Phys. Letters 35A, 391 (1971).

2. D. Salomon, W. Wallner, P. J. West, Mössbauer Effect Methodology 10, 291 (1976).

3. G. Kaindl, D. Salomon, G. Wortmann, Phys. Rev. B8, 1912(1973

4. V. Dornow, J. Binder, A. Heidemann, G. Wortmann to be publ.

5. G. Kaindl, D. Salomon, G. Wortmann, Mössbauer Effect Methodology 8, 211 (1973).

*Pressent address: Argonne National Laboratory, Argonne, Ill. 6043

IMPROVING THE RESOLUTION OF SMALL ENERGY SHIFTS
USING ^{67}Zn and ^{181}Ta

R. V. Pound
Harvard University, Cambridge, Ma. 02178

Mossbauer resonance, as the electromagnetic phenomenon with the smallest known ratio of linewidth to energy, provides a medium for seeking and measuring minute fractional energy shifts that result from external influences. The effect of a change of gravitational potential is probably the most widely known such application. The resonance of ^{57}Fe was almost without competitors for these applications from the exciting time of its discovery[1,2] until very recently. I well remember one evening in early autumn of 1959 perusing with Glen Rebka the tables of radioactive isotopes in search of candidates with which to pursue the application to the gravitational red-shift. We came up with just three examples to try. These were the 0.1 μsec., 14 keV transition of ^{57}Fe, the 9.3 μsec., 93 keV transition of ^{67}Zn and the 4.6 μsec., 13 keV transition of ^{73}Ge. The last of these was, however, recorded as having an internal conversion greater than 1300. Further investigation of the literature revealed that no γ-ray had in fact been detected, which augured badly for detection of resonant absorption. Thus all our efforts were expended on ^{57}Fe[4] and ^{67}Zn[5], the latter, unfortunately, to little avail. In 1961, the 6.8 μsec., 6.25 keV level of ^{181}Ta was reported.[6] This, along with our three, remain the primary media for high resolution applications.

If one desires to detect a small shift of energy by means of γ-ray resonance, one may derive a formulation of the limitation set by statistics in terms of parameters describing the system. In particular, one may consider the intensity transmitted from a source through an absorber as a function of the veclocity of the source relative to the absorber. By observing the intensity at each of two velocities near the points of maximum and minimum slope of the absorption line under conditions producing the shift and in its absence or, preferably, with it reversed, one may calculate an uncertainty of statistical origin in any detected shift in the apparent central resonant frequency in terms of the parameters describing the line. If one assumes the line to be of Lorentzian shape, with a source-absorber velocity $\pm v_H$ required to reduce the absorption to one-half its maximum fraction F, then the uncertainty of the fractional shift may be written

$$\delta a_D = \left[(1 - \tfrac{3}{4} F)/3N_o \right]^{\frac{1}{2}} 8v_H/3Fc \qquad (1)$$

It has been assumed that the observations are made near the velocity of maximum slope of the Lorentz function, i.e., $v_M \approx \pm (\sqrt{3}/3)v_H$; N_o is the number of counts that would have arrived at the detector in the absence of resonant absorption, and c is the velocity of light. In terms of a practical system

$$N_o = R\tau(\epsilon bD)\Omega/4\pi$$

where R is the number of source decays per unit time, τ is the running time, ϵ the detector efficiency, b is the fraction of decays emitting the γ-ray, and D a duty cycle for the velocity modulation system and detector. It is assumed that the detector solid angle Ω is small. For the gravitational red-shift, which increases fractionally with the height h, a fixed detector area causes N_o to decrease as h^{-2}. The net result is that δa_D also increases as h so the statistical resolution of the effect is rendered independent of h. Nevertheless, a large height obviously helps minimize several systematic errors and allows use of strong sources while avoiding high counting rates.

A "figure of merit" for a resonance for such applications can be associated mostly with the smallness of the factor v_H/F, assuming Ω and N_o can be set up equivalently for various sources. Table I summarizes the approximate situation at present for the four most interesting isotopes.

Table I Values of v_H/F in μm/sec.

Isotope	^{57}Fe	^{181}Ta	^{73}Ge	^{67}Zn
Available in Principle	~ 300	~ 12	25	~ 10
Realized Experimentally	~ 300-400	~ 175	125-250	~ 40

In the case of ^{57}Fe, v_H close to the theoretical value of 100 μm/sec. with a fractional depth about 0.3 is achievable. For ^{181}Ta the best result reported is 35 μm/sec. at 0.2 depth,[8] and for ^{73}Ge, 7.8 μm/sec. and 0.06 depth from isotopically enriched single crystal absorber, or 3.7 μm/sec with 0.012 depth with a better natural absorber[9] is reported. For ^{67}Zn line half-widths at the lifetime value of 0.15 μm have been obtained with natural single crystal absorber but with small depth.[10] With high strength from a source diffused into ZnO and an isotopically enriched ZnO absorber, about 0.4 μm/sec. half-width and 1% depth appears to be realizable.[11,12]

In these evaluations, other factors that complicate the employment of these isotopes are not indicated. All three isotopes ^{57}Fe, ^{181}Ta and ^{73}Ge are long lived (270, 140 and 81 days) and the energies result in large recoil free fractions at room temperature. Nevertheless, limitations on absorption depth F occur because of strong non-resonant absorption. In contrast, the 78 hour life and the 93 keV energy that leads to small F even at cryogenic temperatures are strong impediments to applications of ^{67}Zn. Furthermore, although problems of vibration have not been serious in the basic studies where the source and absorber are connected by a rigid transducer system the narrow line would easily be destroyed by vibrations encountered in larger separations especially between independent cryostats. Nevertheless, one may evaluate an example assuming those problems not insurmountable. Suppose that a source of initial strength 1 curie is separated by about 1 meter from an absorber limited in radius, because of the difficulty of obtaining isotopically enriched material, to 1 cm. In such a system, one should be able to obtain 10^{11} useful counts in the life of the source. Application of Eq. (1) then yields a fractional uncertainty of about 6×10^{-19}. Therefore, such a system could, given adequate control of sources of systematic error, determine the gravitational red-shift in the one meter path to about 0.5% for each source, each source experiment lasting about one week and costing perhaps $5,000.

Application of the narrow resonance of ^{73}Ge to such problems remains remote. In spite of the approach to natural linewidth as reported,[9] the basic problem is still the enormous internal conversion coefficient, currently observed to be 1080. As a consequence, the total γ-ray counting rates remain minute and runs as long as three months were utilized to collect the counts that have been used just to define the line shape factors. The need to reject from the data x-rays of far greater intensity militates against compensating for the conversion by increased source strength.

The lines observed so far from ^{181}Ta are of the order of ten times the natural lifetime-limited breadth.[8] Large electric quadrupole interactions as well as electrostatic isomer shifts are encountered and to make sufficiently perfect materials to get closer to natural widths has so far proven difficult. The usual source material is 140 day ^{181}W produced by neutron capture in ^{180}W$(n,\gamma)^{181}$W. Such material has low specific activity compared to material produced by a charged particle beam, ^{181}Ta$(p,n)^{181}$W or ^{181}Ta$(d,2n)^{181}$W. Some effort is being made to see if better source and absorber combinations can be made from the cyclotron produced activity.

A factor that encourages the attention to ^{181}Ta is the fact that the low energy, 6.25 keV, is amenable to being reflected at glancing angles from smooth surfaces, by the technique of total external reflection. Because the atomic electrons are mostly bound weakly compared to 6.25 keV, they behave largely as if free, in response to such an electromagnetic influence. Therefore the index of refraction is, as in the ionosphere for short radio waves, less than one, and the penetrating wave is bent away from the normal to an angle greater than its incidence angle. If incidence is at a sufficiently small angle, no wave propagates in the medium. In glass of ordinary density, the critical angle for 6 keV is about 5 milliradians. For a smooth nickel surface the angle would be about 9 mr. There are strong requirements on the smoothness of the surface even though the short wavelength is projected to a much longer one at the glancing angles of incidence.

W. T. Vetterling and the author have reported elsewhere[13] the tests of this idea of ducting such γ-rays through vacuum pipes with the walls forming a guide through this external reflection. Ordinary glass laboratory tubing of 8 mm bore was carefully aligned, and has been used to duct γ-rays about 7 meters. The γ-rays averaged about four reflections in the length and were attenuated to about 0.50 of the intensity within the capture angle. In addition, an experiment was carried out to show that no change in linewidth or depth accompanied the reflections, in this case using the 14 keV ^{57}Fe resonance through the same tubing. An important property of the "light-pipe" is its low-pass characteristic. For the background γ-rays in a spectrum such as ^{181}Ta, mostly ~56 keV x-rays, the capture solid angle would be only about 0.02 as large as for the 6.2 keV γ-ray and therefore at the end of a pipe that reflected the latter about 8 times there would be far less difficulty resolving the 6 keV line from background.

It is still possible that ways will be found to achieve linewidths in ^{181}Ta closer to the lifetime width than has been achieved so far. One may evaluate the use of this resonance as it has so far been achieved, 35 μm/sec and F ~ 0.2, using Eq. 1. Again we assume a 1 curie source and the solid angle corresponding to nickel coated glass reflectors, a critical angle about 9×10^{-3} radians. The fractional uncertainty for a shift then becomes about 3.5×10^{-17} for a one day run. If a height of 250 meters were used, a gravitational test at the level of 0.0015 (per day)$^{\frac{1}{2}}$ would then be possible. Thus a measurement to about 10^{-4} of the effect could be made in the life of the source. Of course, any progress toward a reduced value of v_H/F would be reflected directly in the results.

The particular application to the gravitational red-shift, although possibly offering about 100-fold improvement over the best experiment so far carried out using γ-ray resonance,[7] may have become academic. Last June 19 a rocket probe carrying a specially designed hydrogen maser was shot out to a height above the surface near 10^4 miles, and allowed to fall back to earth. The experiment and its components were designed and developed under NASA sponsership by R. C. Vessot of the Smithsonian Astrophysical Observatory. Transponder techniques were employed to cancel the enormous first order Doppler shifts and the observed probe-carried clock appears[14] to have given a result in agreement with theory at the level 10^{-4}. The large effort that would be involved in making a new test by γ-ray resonance, although it would still be very small in comparison with the effort expended on the rocket probe and its components, has been rendered harder to muster by the success of that project. Nevertheless, the ultimate sensitivity of the Mossbauer effect as a detector of small shifts remains unchallenged and continued effort to achieve the promised resolution seems well worthwhile.

REFERENCES

1. R. V. Pound and G. A. Rebka, Jr., Phys. Rev. Letters 3, 554 (1959).
2. J. P. Schiffer and W. Marshall, Phys. Rev. Letters 3, 556 (1959).
3. R. V. Pound and G. A. Rebka, Jr., Phys. Rev. Letters 3, 439 (1959).
4. R. V. Pound and G. A. Rebka, Jr., Phys. Rev. Letters 4, 337 (1960).
5. R. V. Pound and G. A. Rebka, Jr., Phys. Rev. Letters 4, 397 (1960).
6. A. H. Muir, Jr., and F. Boehm, Phys. Rev. 122, 1564 (1961).

7. R. V. Pound and J. L. Snider, Phys. Rev. 140, B788 (1965).
8. G. Kaindl, D. Salomon and G. Wortmann, Phys. Rev. B8, 1912 (1973).
9. L. Pfeiffer, Phys. Rev. Letters 38, 862 (1977).
10. W. Potzel, A. Forster and G. M. Kalvius, J. de Physique 37 Colloque C6-427 (1976).
11. G. J. Perlow, W. Potzel, R. M. Kash and H. deWaard, J. Physique 35 Colloque C6-197 (1974).
12. D. Griesinger, R. V. Pound and W. Vetterling, Phys. Rev. B15, 3291 (1977).
13. W. T. Vetterling and R. V. Pound, J. Opt. Soc. Am. 66, 1048 (1976).
14. R. C. Vessot, Accademia Nazionale dei Lincei, Simposio Gravitazione Sperimentale, Pavia, 1976, to be published.

COHERENT NUCLEAR SCATTERING OF SYNCHROTRON RADIATION*

G. T. Trammell and J. P. Hannon
Physics Department, Rice University
Houston, Texas 77001

S. L. Ruby and Paul Flinn
Physics Division, Argonne National Laboratory
Argonne, Illinois

R. L. Mössbauer and F. Parak
Physics Department, Technische Universität München

ABSTRACT

Synchrotron radiation from modern electron storage rings can furnish brighter sources of Mössbauer radiation than radioactive sources if appropriate filtering can be devised. The object is to pass the power in the synchrotron radiation which lies in the spectral range within a few natural widths of the nuclear resonance, with as little diminuition as possible, and to reduce that passed outside this range to a fraction of the filtered radiation.

We consider the methods which can be used to filter, and we make estimates of the efficiency of the various filter schemes.

To achieve this filtering, we must utilize the properties of resonant Mössbauer scattering which distinguish it from the non-resonant electronic scattering. These are: (a) The coherent scattering amplitudes near resonance of low energy nuclear levels are much larger than the nonresonant scattering amplitudes of the atomic electrons even in the forward direction, and furthermore, neglecting polarization factors the nuclear amplitude is isotropic while the electronic amplitude falls off rapidly with scattering angle. (b) The nuclear scattering amplitude is sensitive to the magnetic field direction and to the e.f.g. at the nucleus. (c) The photon polarization dependence of nuclear M1 and E2 scattering amplitudes is quite different from that from atomic electrons. (d) The nuclear scattering amplitudes are strongly frequency dependent while the electronic amplitudes are relatively frequency independent.

1. The simplest filters take advantage of the fact that for thin films the ratio of the resonant to nonresonant Bragg reflectivities are in the ratio of their scattering amplitudes squared. We can then take films which are moderately thick for the resonant fraction so as to give reflectivities of the order of one but still thin for the nonresonant radiation. As an example the scattering amplitude at resonance for the Fe^{57} 14.4 kev photon is equivalent to ∿440 electrons while the electronic scattering amplitude corresponds to 26 electrons in the forward direction and to only 7.6 electrons at

*Work supported in part by the National Science Foundation.

90° scattering angle. The (332) Bragg reflection in Fe corresponds to very near 90° scattering angle. If we consider a thin film consisting of M such layers then the scattered flux near the Bragg angle is

$$P(\omega) \doteq I_o \Delta \left[\frac{(F_o M)^2}{1+(\Delta\omega/\Gamma/2)^2} + (F_e M)^2 \frac{\sin^2(M\pi \frac{\Delta\omega}{\omega_o})}{(M\pi \frac{\Delta\omega}{\omega_o})^2} \right]$$

where I_o ($cm^{-2}ev^{-1}millirad^{-1}sec^{-1}$) is the incident flux, Δ is the beam collimation which we have taken $\Delta \overset{<}{\sim} M^{-1}$, ω_o is the resonance energy, $F_o \doteq 10^{-3}$ is the scattering amplitude per plane at resonance, while $F_e \doteq 2 \times 10^{-5}$ is the electronic planar scattering amplitude. A thin film with $M = 10^3$ layers is already moderately thick for the resonance portion while still very thin for the nonresonant radiation. (For $M = 10^3$ the first term in the square brackets is $\doteq 0.25$ at $\Delta\omega = 0$, rather than one as indicated, due to multiple scattering.) The reflected power in the resonance region is $P_N = I_o \cdot \Delta \cdot \frac{\pi\Gamma}{2} (F_o M)^2$ while for the nonresonance part it is $P_e = I_o \cdot \Delta \cdot \frac{\omega_o}{M} \cdot (F_e M)^2$. Taking $M = 10^3$ the reflection coefficient in the spectral region $\Delta\omega \sim \frac{\pi\Gamma}{2}$ is $(F_o M)^2 \sim 1$, whereas the reflection coefficient in the spectral region $\Delta\Omega \sim \omega_o \times 10^{-3}$ is $(F_e M)^2 \sim 4 \times 10^{-4}$. Now $10^{-3} \omega_o/\Gamma \sim 3 \times 10^9$, therefore it would require three 90° reflections from parallel 10^3 layers thick crystals to achieve the desired filtering in this simple system. As mentioned above, the reflection coefficient at resonance is ~ 0.25 rather than $(F_o M)^2 = 1$. Therefore the reflected resonance flux after three thin film reflections would be

$$P_N \doteq (I_o \cdot \Delta \cdot \frac{\pi\Gamma}{2})/64.$$

2. Better thin film filters can be obtained using the other distinctive differences listed: i) The synchrotron radiation is polarized with $P = \frac{I_\perp}{I_{||}} \doteq 10^{-1}$ where $I_{||}$ and $I_\perp$ are the intensities $||$ and $\perp$ to the synchrotron plane. By double 90° scattering from a good Si channel cut crystal we estimate that we can produce a beam with $\frac{I_\perp}{I_{||}} < 10^{-7}$, with $I_{||} \doteq 1/3 I_o$, and with $\Delta\Omega \doteq 10^{-4} \omega_o$. If this radiation is then 90° reflected from a thin film of Fe^{57} with the polarization in the scattering plane then the electronic scattering amplitude is reduced by a factor $\sim \frac{\hbar\omega_o}{mc^2}$ from that given

48

in (1) and $(F_e M)^2 \doteq 10^{-7}$ for $M = 10^3$, while F_N remains unchanged since the nuclear scattering is M1. In this case after a single scattering $P_e/P_N \doteq 30$, and $P_N \doteq \frac{1}{12} (I_0 \Delta \cdot \frac{\pi \Gamma}{2})$. Although the total nonresonance power passed by this filter $\doteq 30$ times that in the Mössbauer slice the reduction of $P(\omega)$ by a factor of 10^7 off resonance would suffice to make this a good beam for most Mössbauer experiments. ii) A thin crystal grown by epitaxially depositing alternate layers of Fe^{57} and Fe^{56} could serve as a filter when set for a superlattice Bragg reflection. The off-resonance reflectivity would be $F_e^2 \doteq 10^{-9}$. For $M \doteq 10^3$ this would be an excellent filter except for the higher order reflections: Although the 14.4 kev region would be very pure the reflected power at 28.8 kev would be $\sim 10^5 \frac{I_0(28.8)}{I_0(14.4)}$ larger and might give an intolerable background even with energy sensitive detectors. The higher order reflections would similarly present problems for thin film anti-ferromagnetic reflectors or from materials with zero chemical structure factors for some reflections.

3. <u>Time filtering.</u>[1] The synchrotron pulses are very short $\tau \gtrsim 2 \times 10^{-10}$ sec (SPEAR) with pulse separation $\Delta \gtrsim 10^{-6}$ sec (SPEAR). The nonresonant electronic processes (scattering, photo and auger electrons) occur essentially instantaneously over the duration of the pulse, whereas the nuclear resonance processes come out with a mean time delay Γ^{-1}. For most Mössbauer resonances $\tau < \Gamma^{-1} < \Delta$, so a timed detector which can be restored after the initial prompt pulse in a time short compared to Γ^{-1} can measure the resonance processes by temporal separation from the nonresonance processes. An initial filter of the sort discussed in (2) could be used to avoid overloading the detector during the prompt pulse.

An interesting aspect here is that there will be beats in the resonantly scattered γ's and conversion electrons if there are hyperfine splittings.[2]

4. The thin crystal filters will give a Mössbauer line which is Lorentzian with a width on the order of the natural line width. For some purposes however thick filters are advantageous. The Bragg reflectivities are several times higher (0.87 (thick crystal) vs. 0.25 ($M = 10^3$) for the 90° (332) reflections in Fe^{57}), and the resonantly filtered radiation has a greatly broadened width ($\sim 50\Gamma$ for Fe^{57}), so that the total reflected power is much increased. Although the greatly broadened width makes the filtered beam not so suitable for Mössbauer work, it still has a very long coherence length (~ 1 m vs. $\sim 1 \mu$m for "monochromatized" X-rays or the output

of an X-ray laser), and is ideal for interferometry applications.

The effects of heating and photo-electron emission are more severe for the thick crystal filters, and thin crystal filters can be operated in considerably higher synchrotron fluxes. However, for current synchrotron sources, and those planned for the immediate future, the photon fluxes are sufficiently low that these effects should pose no problems.

A detailed discussion of all these considerations will be presented elsewhere.[3]

REFERENCES

1. S. L. Ruby, Journal de Phys. C6, 209 (1974).
2. G. T. Trammell and J. P. Hannon, to be submitted to Physics Letters.
3. G. T. Trammell, J. P. Hannon, S. L. Ruby, P. Flinn, R. L. Mössbauer, and F. Parak, to be submitted.

Symmetric Radiant State in Nuclear Resonant Bragg Scattering

S. L. Ruby

Argonne National Laboratory, Argonne, Illinois 60439*

The purpose of this paper is to describe a qualitative way to understand the temporal behaviour of nuclear resonant Bragg scattering. If you need quantitative detailed analysis, then use the careful work of Hannon and Trammell[1] who give scattering amplitudes from crystals for monochromatic photons, fold in the incoming wave packet frequency distribution, and Fourier-transform into the time domain. The present analysis is very crude, but is helpful in reaching an 'intuitive' understanding of a subject that has generated its share of confusion.

In the usual non-resonant scattering of x-rays from the regularly spaced electrons of a crystal, there has been little need to explicate the temporal behaviour. Such scattering is treated using the time-independent scattering amplitudes of the atom for precise frequencies. The atoms can be combined into lines, or planes, or crystals, and the corresponding amplitudes developed. Where the scattering amplitudes are rapidly changing functions of frequency, as is the case for nuclear scattering from a crystal, one anticipates delays of about the nuclear lifetime. Such delays of about 100 nanoseconds have been seen in both incoherent[2] and coherent[3,4] scattering experiments.

The scattering theory has been carefully applied to the case of an enriched non-magnetic Fe crystal illuminated on the Bragg angle with nuclear resonant light, and it has been calculated that the reflectivity of the crystal remains large for perhaps fifty line widths.[1] In the near future, the explicit time behaviour will be calculated since experiments of this kind are in development. The broad reflectivity suggests an enhanced radiative line width, but it is not easy to give an intuitive interpretation of its origin. The power densities in the problem clearly rule out stimulated emission. Trammell[5] first showed that the excess line width increases with the number of planes. Zaretski and Lomonosov[6] introduced the idea of the nuclear exciton, whose decay is enhanced and directional. Kagan[7] has noted that on Bragg the internal conversion decay channel is weakened as if α is decreasing, and calls this 'suppression of the inelastic channel'. We wish to show the intermediate state in Bragg scattering is a 'super-radiant' symmetric state,[8] and thus explain its rapid decay.

Simplify the crystal to N_x, N_y, and N_p nuclei in the x, y, and z directions with spacings a, b, and d. Assume 100% enrichment, suppress polarization directions, and fix the nuclei rigidly - no thermal vibration. In addition keep the density of nuclei in the

*Work performed under the auspices of the USERDA, Division of Physical Research.

planes, and the number of planes, such that there is not appreciable absorption (Born approximation). To allow for time delay
experiments, this light is chopped much shorter than the nuclear
lifetime: the wave packet thus has an energy spread much wider
than the nuclear width. However, for an approximate calculation,
we suppress the detailed dependence on energy, and ψ refers to
all frequencies in the wave packet. And the incident power is
low, so that even a single excitation in the crystal is infrequent.
Such an excitation, or intermediate wave function, can be written
as

$$\psi_{int} = (N_x N_y N_p)^{-\frac{1}{2}} \sum_{\ell}^{N_x} \sum_{m}^{N_y} \sum_{n}^{N_p} e^{i \vec{k}_e \cdot \vec{r}_{\ell mn}} |e_{\ell mn} g_s> \qquad (1)$$

Clearly the relative phases of each ket merely represents the
arrival time of the plane wave at the various positions. Otherwise
each nucleus is treated alike. It is the single excitation, three
dimensional version of Dicke's symmetric state. If the $r_{\ell mn}$ are
random, as in a glass, the behaviour of ψ_{int} will be incoherent or
random, despite the symmetry of (1). Note that even in a regular
crystal, if the excited state were created by such localized
effects as β-decay or Coulomb excitation, ψ_{int} would not be the
appropriate description.

What is the radiative decay behaviour of ψ_{int} ? It will
decay to a final state $|g_1 g_2 \ldots g_n> = |G>$ with a photon $\vec{k}_f$.
Putting $<G|O_r|e_{\ell mn} g_s> = M_o$, the matrix element is easily obtained.

$$M = M_o (N_x N_y N_p)^{-\frac{1}{2}} \sum_{\ell mn} e^{i(\vec{k}_i - \vec{k}_f) \cdot \vec{r}_{\ell mn}} \qquad (2)$$

which is exactly the usual x-ray result. Following traditional
paths now, let $\vec{k}_i = k(\hat{1} \cos \theta i - \hat{k} \sin \theta i)$, $\vec{k}_f = k(\hat{1} \cos \theta + \hat{j} \sin \theta \sin \psi + \hat{k} \sin \theta \cos \psi)$, $\vec{r} = \hat{1} \ell a + \hat{j} mb + \hat{k} nd$, and calculate
M^*M.

$$M^*M(\theta,\psi) = \frac{M_o^2}{N_x N_y N_p} \sum_{\ell mn} \sum_{\ell' m' n'} e^{i(\vec{k}_i - \vec{k}_f) \cdot (\vec{r}_{\ell mn} - \vec{r}_{\ell' m' n'})} \qquad (3)$$

which leads to

$$M^*M(\theta,\psi) = M_o^2 \left[1 + \frac{1}{N_x N_y N_p} \; A \cdot B \cdot C \right] \tag{4}$$

where $A = \dfrac{\sin^2 N_x \alpha}{\sin^2 \alpha} - N_x$ with $\alpha = \dfrac{ka}{2}(\cos \theta i - \cos \theta)$

$B = \dfrac{\sin^2 N_y \beta}{\sin^2 \beta} - N_y$ with $\beta = \dfrac{kb}{2}(-\sin \theta \sin \psi)$

$C = \dfrac{\sin^2 N_p \gamma}{\sin^2 \gamma} - N_p$ with $\gamma = \dfrac{kd}{2}(\sin \theta i + \sin \theta \cos \psi)$

It is easily shown that M^*M has only two narrow exit channels – the transmitted and reflected beam. If eq. (1) had a random phase in each term, as is appropriate for β-decay, then M^*M would be independent of direction. By integrating the intensity M^*M over all directions, the radiative line width γ_r is found.

Note that $\int_o^\infty d\varepsilon \dfrac{\sin^2 N_z \varepsilon}{\sin^2 z\varepsilon} = \dfrac{\pi}{2} \dfrac{N}{z}$. Also the result is specialized

to non-zero values of θ_B.

$$\gamma_r = \gamma_o \left[1 + \frac{2\pi}{k^2 ab \sin \theta_B} \; N_p \right] \tag{5}$$

This shows that the intermediate Bragg state decays with an enhanced radiative width, or shorter lifetime, compared to an isolated excited nucleus.

For ^{57}Fe using $\theta_B = 45^o$, $\gamma_r = \gamma_o(1 + .03 N_p)$. The observed lifetime includes the unchanged incoherent internal conversion process whose coefficient $\alpha = 8.4$, and therefore the overall width γ corresponds to $\gamma = \gamma_o(1 + .03 N_p) + 8.4 \gamma_o = \gamma_o(9.4 + .03 Np)$. For $N_p \cong 10^3$, one expects $T \simeq 25$ nanoseconds, not the ordinary 100. Clearly, the 30-fold decrease in the radiative lifetime, from 1000 nsec to about 30 nsec, greatly reduces the fraction of times that the intermediate state can decay via the 110 nsec internal conversion channel.

It has been shown above, by direct calculation, that a Bragg intermediate state (ψ_{int} with special $\vec{r}$) decays differently than the more usual 'random' intermediate states. The differences are two sharp Bragg peaks in direction versus isotropic decay, and decreased lifetime. Can this surprising fact be made even simpler? Independently both Dicke[8] & Trammell[5] have discussed a two body system, each of which can be in the ground (g) or excited (e) internal state. They are labeled 1 and 2 on the basis of position, even though ka < < 1 with k the reduced wavelength and a the separation distance. Now kets like $e^{i\phi_1}|e_1g_2>$ or $e^{i\phi_2}|g_1e_2>$ describe situations where a particular one of the atoms is excited. To describe that only one nucleus is excited, but not knowing which, is done by

$$\psi \text{ random} = \frac{1}{\sqrt{2}}\left\{e^{i\phi_1}|e_1g_2> + e^{i\phi_2}|g_1e_2>\right\}.$$

Till now the phases ϕ_1 and ϕ_2 have been random, but if the excitation has been such that there is a definite phase difference ϕ between particles 1 and 2, then one needs

$$\psi = 2^{-\frac{1}{2}} \cdot e^{i\phi_1}\left\{|e_1g_2> + e^{i\phi}|g_1e_2>\right\}$$

and this state is most easily described as a linear combination of the symmetric

$$\psi_S = \frac{1}{\sqrt{2}}\left\{|e_1g_2> + |g_1e_2>\right\}$$

and the anti-symmetric eigenvectors.

$$\psi_A = \frac{1}{\sqrt{2}}\left\{|e_1g_2> - |g_1e_2>\right\}.$$

The important point that $\gamma_S = 2\gamma_o$ while $\gamma_A = 0$, is easily shown.

$$\left[\gamma = \int d\Omega| \, 2^{-\frac{1}{2}} < |e_1g_2> \pm |g_1e_2> \, |0_r| \, |G \, k_f>|^2 \right.$$

$$\left. = \int d\Omega \, \frac{M_o}{2} \, | \, 1 \pm 1 \, |^2 = \gamma_o(1 \pm 1)\right].$$

Note that the random intermediate state can be described as a linear combination of ψ_S and ψ_A, and will have $\gamma = \gamma_o$. If this is generalized to n atoms instead of two, $\gamma_S = n \gamma_o$ while all the (n – 1) anti-symmetric states have zero decay rates. If you can prepare symmetric states, they decay rapidly!! In electronical engineering language, to give the classical version, two close dipoles decay between four or zero times more rapidly than one, depending on the phase difference. So, at the end, the surprising (?) properties of enhanced, directional decay for nuclei become the old-fashioned properties of a phased array of antennas.

It is worth noting again that just having a high density of scatterers and an exciton is not enough for enhancement. For example, moving θ away from θ_B by a fraction of a degree will kill off the enhancement completely. Generally, enhanced decay rates are not found merely when more scatterers are used, for the

54

amplitudes add constructively in some regions of space, destructively in others, and overall the integrated intensity stays fixed. Only if the amplitudes add constructively every where in space will line widths increase.

Note that in the above description, the photon occupation number is zero when we calculate γ_r. The above symmetric radiant state has been specialized to only one excited nucleus. Dicke has considered super-radiant states with n atoms, of which m are excited (m < n). The temporal evolution of such states can involve stimulated emission, but not for our case.

Such states are not confined to γ-ray optics, of course. Analagous behaviour takes place in neutron scattering, and perhaps also with visible light.

A final 'intuitive' description - the incident plane wave interacts with many nuclei at once. Except at the Bragg angle, there is only incoherent summing of amplitudes, and the results are as if each nucleus scatters individually. At Bragg, the increased amplitudes for radiative scattering coming from the coherent symmetric addition leads to enhanced radiation widths for both absorption and reemission. Since internal conversion cannot be coherent with the final state marked by a K-hole, and is not speeded up, it is observed less frequently as T_{rad} decreases. It is more insightful to talk, not of suppression of the inelastic channel, but enhancement of the radiative channel.

I wish to acknowledge the tolerance of many colleagues as I tried to clarify these ideas. In particular, J. Hannon, M. Peshkin, P. Flinn, H. Lipkin, and R. Ringo all helped eliminate many incorrect earlier versions.

REFERENCES

1. J. P. Hannon and G. T. Trammell, Phys. Rev. 186, 306 (1969).
2. P. Thieberger, J. A. Moragues, and A. W. Sunyar, Phys. Rev. 171, 425 (1968).
3. F. J. Lynch, R. E. Holland, and M. Hammermesh, Phys. Rev. 120, 513 (1960).
4. S. Bernstein and E. C. Campbell, Phys. Rev. 132, 1625 (1963).
5. G. T. Trammell, Chapter in "Chemical Effects of Nuclear Transformations", IAEA, Vienna (1961).
6. D. F. Zaretski and V. V. Lomonosov, JEPT, 21, 243 (1965).
7. Y. Kagan and A. M. Afanaséev, JETP, 49, 1504 (1965).
8. R. H. Dicke, Phys. Rev. 93, 99 (1953).

CALIBRATION CONSTANTS IN MÖSSBAUER DIFFRACTION EXPERIMENTS*

J. G. Mullen and J. R. Stevenson
Physics Department, Purdue University, West Lafayette, IN 47907

ABSTRACT

We have measured the ratio of the inelastic to total scattering for the (200) Bragg peak of LiF to be $(1.4 \pm 1.0)\%$. This result is an order of magnitude less than reported by O'Connor and Butt[1], and it can be used as a basis for a simpler calibration of the elastic and inelastic components of scattering for other Mössbauer diffraction experiments.

INTRODUCTION

In the early work of O'Connor and Butt[1] it was reported that the room temperature fraction of inelastic to total scattering for the (200) Bragg reflection from LiF is about 20%. Since this pioneering work in 1963, there have been several papers[2-9] on Mössbauer Diffraction experiments using the relatively ideal Fe^{57} 14.4 keV line. In these experiments the partitioning of the scattered x-rays into elastic and inelastic parts depends critically on a calibration experiment. Two schemes have thus far been used for calibrating the elastic and inelastic components and both are fought with experimental difficulties and hard to access systematic errors.

Here we describe a new procedure based on comparing the on-and off-resonance radiation for a system under study with that of the (200) Bragg peak of LiF.

EXPERIMENTAL METHODS

In Figure 1 we show schematically the two common approaches for calibrating Mössbauer diffraction experiments, along with the present approach using the (200) Bragg peak of LiF.

The direct beam technique (DBT) involves comparing on-and off-resonance scattering (Figure 1a) for the Bragg peak being studied with that of the direct beam undeflected by an intermediary material. The fractional effect, P, given by

$$ P = \frac{I_\infty - I_o}{I_\infty} , \qquad (1) $$

where I_∞ is the background corrected off-resonance intensity and I_o is the background corrected on-resonance intensity, gives the calibration partitioning of the elastic and inelastic components of the scattered beam. Thus, $P(\theta)$ for the scattered beam compared to $P(0)$

*Work supported by the National Science Foundation contract No. DMR 76-00889.

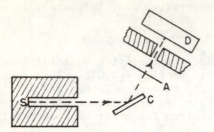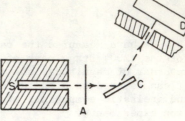

a)   b)

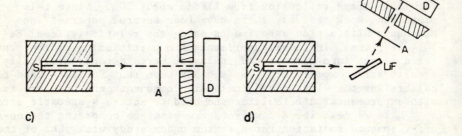

c)   d)

Figure 1 - Schematic of Techniques for Measuring Calibration Constant, S-source, A-absorber, D-detector, C-crystal being studied. LiF - lithium fluoride single crystal.

for the direct beam gives the fraction of the scattered beam which
is elastic. In many experiments, the inelastic component is small
which means that $P(O)$ and $P(\theta)$ must be very accurately determined
for at least one point, which requires a very accurate background
measurement. Because of the great intensity of the 122 keV radiation, its background contribution is generally found by a series of
measurements with graded thicknesses of aluminum, all thick enough
to block the transmission of the 14.4 keV line. This measurement is
very time consuming, particularly for the scattered beam, and is
complicated by a dubious extrapolation procedure to find the background contribution in the zero thickness limit. Also, the count
rates are so high that it is typically necessary to make large (30%)
dead time corrections, or, alternatively to alter the geometry which
brings in added systematic errors.

In the shifted absorber technique (SAT) the high count rates of
the DBT are avoided by scattering from the crystal being studied, as
is illustrated in Figure 1b. It is still necessary to correct for
the 122 keV x-rays and our data suggests that the log $\sim$(intensity)
vs. thickness is not linear, which is not unexpected, since down
scatter can occur from both the aluminum absorbers as well as the
crystal under study. Low count rates again makes this calibration
lengthy and tedious. Also, because these experiments require highly
divergent beams ($\sim 2^\circ$) it is essential that the absorber used be extremely uniform, as the solid angle subtended between the source and
absorber changes when the absorber is shifted, as can be seen from
an examination of Figures 1a and 1b.

RESULTS

Using a mechanical drive system, with a Co^{57}:Pd source and a
stainless steel absorber, we have measured the ratio of the inelastic to total scattering on the (200) Bragg peak of LiF by both DBT
and SAT. Using SAT we find

$$\frac{I_{inel}}{I_{total}} = (1.9_6 \pm 1.0)\% \quad ,$$

while with DBT we find

$$\frac{I_{inel}}{I_{total}} = (0.0 \pm 1.7)\% \quad .$$

The weighted mean value of these measurements gives

$$\frac{I_{inel}}{I_{total}} = (1._4 \pm 1.0)\% \quad .$$

We have calculated the total inelastic TDS by the method of Cooper
and Rouse[10] and find it to be about 0.2%.

Our observation of nearly perfectly elastic scattering from the LiF (200) Bragg peak is important in that it makes possible a new and simpler method for calibrating Mössbauer diffraction experiments for any crystal. By comparing P for the crystal Bragg peak being investigated (Figure 1a) with P for the (200) Bragg peak of LiF (Figure 1d) we can get a measure of elastic and inelastic components of the diffraction line. For approximate results it will not be necessary to make the difficult graded aluminum background correc tion, which is only of order 1%, for the LiF (200) Bragg peak. Also the LiF technique will approximately half the counting time for weak reflections compared to SAT, which requires double counting for each point where the elastic and inelastic components are determined.

High accuracy with the present approach will require correcting for the TDS and Compton scattering, although the Compton component will necessitate a separate measurement away from the region of the Bragg peak and will be dependent on the geometry of the collimating slit system. The total TDS can be found by the programs of Cooper and Rouse[10] which gives 0.2%. For typical geometries this will be a reasonable estimate of the required TDS component.

REFERENCES

1. D. A. O'Connor and N. M. Butt, Phys. Letters 7, 233 (1963).
2. C. Ghezzi, A. Merlini, and S. Pace, Nuovo Cimento 64B, 103 (1969).
3. G. Albanese, C. Ghezzi, A. Merlini, and S. Pace, Phys. Rev. B5, 1746 (1972).
4. G. Albanese, C. Ghezzi and A. Merlini, Phys. Rev. B7, 65 (1973).
5. N. M. Butt and G. Solt, Acta Crystal. A27, 238 (1971).
6. G. Albanese and C. Ghezzi, Phys. Rev. B8, 1315 (1973).
7. D. A. O'Connor and E. R. Spicer, Phys. Lett. 29A, 1361 (1969).
8. B. W. Battermann, G. Maracci, A. Merlini, and S. Pace, Phys. Rev. Lett. 31, 227 (1973).
9. G. Albanese and C. Ghezzi, Phys. Stat. Sol. (a) 22, 209 (1974).
10. M. J. Cooper and K. D. Rouse, Acta Cryst. A24, 4051 (1968).

THE SELECTIVE EXCITATION METHOD

G.R. Hoy*

DRF/CPN/Groupe Interactions Hyperfines

CEN-Grenoble, 85 X-38041 Grenoble, France

ABSTRACT

Recently progress has been made in understanding the selective excitation and subsequent decay of nuclear energy levels where in both events, the Mössbauer effect occurs. The spectral distribution of the emitted radiation in the absence of time dependent hyperfine interactions seems well established. The situation is much more complicated if one includes time dependent effects. However, there has been considerable progress made in this case also. The general technique has already been applied to study the nature of the Morin transition in αFe_2O_3 and relaxation effects in magnetite (Fe_3O_4). Further theoretical and experimental results are necessary in order to properly evaluate to power of the method.

INTRODUCTION

One of the most intriguing aspects of the selective excitation method[1] is in the study of relaxation effects. In ordinary Mössbauer transmission spectroscopy, relaxation processes must often be identified by spectral line broadening. However, other factors, notably inhomogeneous hyperfine fields, will also lead to line broadening. Therefore, it is not unusual for such studies to be somewhat in error. This new method of selective excitation may prove to be very powerful, but more research is needed. The basic idea of the technique is to have a single line source move at constant velocity such that the emitted Mössbauer photons have the

* Permanent address : Physics Dept. Boston University, Boston, Massachusetts (USA). Supported by the National Science Foundation, Grant No. DMR 73-07665A03.

correct energy to excite a particular nuclear level in the polycrystalline material under study (called the scatterer). In this way an excited nuclear ensemble is carefully prepared, and the subsequent nuclear recoilless gamma ray de-excitation can be energy analyzed by using a single line Mössbauer absorber (called the analyzer) driven in the constant acceleration mode chosen to correspond to the energy range of interest. If during the lifetime of the excited nuclear state there are hyperfine interactions which perturb the prepared ensemble, the energy spectral distribution of the recoillessly emitted gamma rays will be different from that obtained if the ensemble were unperturbed. An example of such a perturbing hyperfine interaction would be when the effective magnetic field at the nucleus is flipping between "up" and "down" orientations.

BASIC CONSIDERATIONS

Before considering the more complicated case of relaxation effects, it was necessary to demonstrate that the technique itself worked. This was shown in 1968-1969[2], [3], [4]. However, in these early works there was no attempt made to actually calculate the expected result. Approximate expressions for the energy distributions of the Mössbauer scattered radiation including thickness effects in the scatterer (made of the material under study), but neglecting interference were published[5] in 1974. These calculations were in quantitative agreement with the experimental results, and thus give a basis for understanding selective excitation experiments in the absence of relaxation effects. One of the most important results of those calculations[5] was that the energy distribution of the emitted radiation has its maximum at the value of the incoming (incident) radiation. This results from the fact that the scatterer must be thick (β or $T_a \sim 300-400$) in order to perform the experiment in a reasonable length of time. Thus even when a particular nuclear transition is excited off resonance, the emitted radiation appears at the excitation energy. A further consequence of this result is that inhomogeneous fields contribute to the selective excitation spectrum in a way that is different from that due to relaxation effects. This feature of the technique is not generally appreciated.

In general, Rayleigh scattering i. e. elastic scattering of the incident beam by the atoms in the scatterer, is coherent with nuclear resonant scattering. However, for a scattering angle of 90° the interference term[6] cancels and one can simply add the intensities. When using polycrystalline scatterers and only the 90° scattering geometry, Bragg scattering just produces a background term.

APPLICATION TO TIME DEPENDENT PROBLEMS

As a first step in looking for time dependent effects, one can look for selective excitation experimental results that differ from the time independent calculation given in reference 5. Ultimately one needs a general calculation valid for all relaxation rates and all energy values of excitation. The basic theoretical problem of relaxation effects in such scattering experiments has been solved[7,8,9]. However, these theoretical results need to be calculated including the incident beam line shape, thickness effects in the scatterer, and Rayleigh scattering, so that a direct comparison can be made with experimental results. Until now, the only such experiments using polycrystalline scatterers, have been to study the nature of the Morin transition[5] in $\alpha Fe_2 O_3$ (hematite) and relaxation effects[10] in $Fe_3 O_4$ (magnetite). The results of the hematite experiments indicate that the Morin transition does have a dynamic character. The theoretical analysis[5] was only approximate, and the model not completely realistic, thus the relaxation times found are only order of magnitude values. The magnetite study was initiated because of the rather surprisingly slow value found for the electron hopping time at the B sites ($\sim 10^{-9}$ sec.) at room temperature from ordinary Mössbauer measurements[11]. As mentioned in the introduction such experiments are not always unambiguous. The selective excitation results show that at room temperature the electron hopping is much faster than 10^{-9} sec.[10] and, in fact, other recent Mössbauer measurements[12] show that the broadening originally observed was due to inhomogeneous fields. A further result of the selective excitation work on magnetite was that below the Verwey transition (119 K), there are dynamical effects probably occurring on a time scale of approximately 10^{-7} sec.

ACKNOWLEDGEMENT

The author would like to acknowledge the efforts of Dr. Bohdan Balko, who as a graduate student enthousiastically accepted this difficult problem. Without his hard work much of the work quoted above would not have been possible.

REFERENCES

1 For a review of the technique sec B. Balko and G.R. Hoy,
"Mössbauer Effect Methodology", Vol. 9, edited by L.G. Gruverman
(Plenum, New York 1974).

2 A.N. Artemev, G.V. Smiroy, and E.P. Stepanoy,
Sov. Phys.-JETP 27, 547 (1968).

3 W. Meisel, Monatsber, Dtsch. Akad. Wiss. Berlin Ger. 11, 355
(1969).

4 N.D. Heiman, J.C. Walker, and L. Pfeiffer, Phys. Rev. 184, 281
(1969).

5 B. Balko and G.R. Hoy, Phys. Rev. B10, 36 (1974).

6 P.J. Black, D.E. Evans, and D.A. O'Connor, Proc. R. Soc. Lond.
A270, 168 (1962).

7 A.M. Afanasev and V.D. Gorobchenko, Sov. Phys.-JETP 40, 1114
(1974).

8 F. Hartmann-Boutron, J. Physique 37, 549 (1976).

9 M. Blume, private communication.

10 B. Balko and G.R. Hoy, J. Physique Colloq. (C6) 37 C6-89 (1976).

11 W. Kundig and R.S. Hargrove, Solid State Commun. 7 223 (1969).

12 A.M. van Diepen, International Conf. on Magnetism,
Sept. 6-10, Amsterdam (The Netherlands) 1976.

COHERENCE AND INTERFERENCE IN THE MÖSSBAUER EFFECT

Harry J. Lipkin†
Argonne National Laboratory, Argonne, Illinois 60439*
and
Fermi National Accelerator Laboratory, Batavia, Illinois 60510

When asked to give this talk I was told that it should be an after dinner talk and not a regular talk for the meeting. But when I asked if it should be understandable to the wives present at the dinner the answer was "Oh no, it should be on a high, professional level". So no matter which way I pitch this talk, it will be criticized from both directions.

At this point it is worth recalling some of the early history of the Mössbauer effect. I got into the Mössbauer business in 1958 while spending a year at the University of Illinois in Urbana directing Hans Frauenfelder's group doing parity non-conservation experiments in beta decay while Hans was on sabbatical at CERN. It was becoming evident that all the interesting parity experiments had been done and it was time to look for new directions. Hans ran across a peculiar work by a young German physicist named Mössbauer and sent us the reprint with the suggestion that we look into it and see whether it was worth pursuing in Urbana. One glance at the paper was sufficient to see that I couldn't understand it because it involved some solid-state physics which I did not know. But since Urbana is one place where one should be able to find out everything necessary in solid-state physics, we went to Fred Seitz and asked him to explain this paper to us.

Seitz glanced at the paper and then asked "Who is this fellow Mössbauer?" "Does anyone know him?" "Is he reliable?". He then asked for a few days to look it over and think about it. Shortly afterward he caught me and said "I have looked at this Mössbauer paper and it is perfectly all right. But I must admit that when I first looked at it I thought it was completely crazy".

This was the beginning of a study in complementarity coherence and interference which plagues all efforts to understand the Mössbauer effect. In addition to the usual standard quantum-mechanical effects, there is an additional complementarity effect. We are used to the relation

$$\Delta p \cdot \Delta x \sim \hbar \qquad (1a)$$

describing complementarity in the description of the electron. If

*Work performed under the auspices of USERDA, Div. of Phy. Research.
†On leave from Weizmann Institute of Science, Rehovot, Israel.

the electron is described in a way which looks sharp to a "position measuring" physicist, it looks fuzzy to a "momentum measuring" physicist. It it looks sharp to a "momentum measuring" physicist, it looks fuzzy to a "position measuring" physicist.

The new complementarity in the description of the Mössbauer effect is described by the equation

$$\Delta N \cdot \Delta S \sim \hbar. \tag{1b}$$

If the Mössbauer effect is described in a way which looks sharp to a nuclear physicist it looks fuzzy to a solid-state physicist. If it looks sharp to a solid-state physicist, it looks fuzzy to a nuclear physicist.

The problem with understanding the Mössbauer effect in the early days was that it required a knowledge of a few elementary things in both nuclear and solid-state physics but it was very hard to find any one physicist who knew these elementary things in both areas. As a nuclear physicist in Urbana I managed to learn the necessary elements of solid-state physics to become an expert on solid-state for the nuclear physicists while at the same time remaining an expert in nuclear physics who could now talk solid-state language.

One subtle point which confused Fred Seitz and all solid-state physicists was the difference between the energy of the gamma ray, the momentum of the gamma ray and the energy of the recoil nucleus. When a solid-state physicist hears that a nucleus emits a gamma ray with an energy of over 100 kV, he knows that 100 kV is so much greater than all lattice energies that it is nonsense to talk about the energy and momentum transfer through the crystal in any simple way. However, the _energy_ of 100 kV leaves the lattice. When a _momentum_ transfer of 100 kV is given to an iridium nucleus, it imparts a kinetic energy which is only of the order of thermal energies and the same order as lattice energies. Thus, the 100 kV of photon energy is a red herring. As soon as the solid-state physicist sees the recoil in terms of the kinetic energy of the recoil nucleus, he is back on familiar ground very well investigated in problems of neutron capture and neutron scattering in crystals.

After learning enough solid-state physics to think in a sensible way about the Mössbauer effect, I wrote a paper explaining it in a simple way and published it in Annals of Physics. This may be the first paper which used the term Mössbauer effect in the title and possibly used it at all. This is because at that time in the Fall of 1959, the physics community had not yet come to the conclusion that Mössbauer's work was sufficiently important to have the effect named after him. At first, the physics community did not believe the experiments. Then after they velieved them they thought it was a peculiar effect which would not really have any serious consequences. Both nuclear and solid-state physicists claimed there was nothing that you could do with the Mössbauer effect that you couldn't do better with other techniques, and it would remain an educational curiosity. Today we see how much everything has changed since those early days.

The Mössbauer effect provides an interesting example of the unity of physics, as opposed to the specialization which is seen almost everywhere else. It requires a combination of different points of view for an understanding of the effect, and its applications cover a very broad area, as can be seen at this conference. Another interesting aspect of the effect is that very similar phenomena occur in many other areas and the similarity is generally not recognized.

The basic principle of a "recoil-free" transition occurs in many areas in physics, but it is usually not called by that name, and the relation between such transitions in different areas is not usually recognized.

Consider first Bragg scattering, which was known long before the Mössbauer effect. When an x-ray is scattered by different atoms in a crystal, why are the scattered waves coherent? The momentum of the scattered x-ray photon is different from the momentum of the incident x-ray photon, so there must be a momentum transfer to the atom which scattered the x-ray. A momentum measurement would determine which atom has scattered the photon, so the scattered waves from different atoms are not coherent.

But when Bragg scattering occurs, a momentum measurement cannot determine which atom has scattered the photon. The uncertainty principle (1a) tells us that the momentum of the atom has an uncertainty associated with the uncertainty in its position. For any practical experiment, the uncertainty Δx in the position of the atom must be small compared to the wave length of the radiation,

$$\Delta x \ll \lambdabar \qquad (2a)$$

From this and the uncertainty principle (1a) the uncertainty in momentum is

$$\Delta p > \hbar/\Delta x \gg \hbar/\lambda \sim p \qquad (2b)$$

where p is the momentum of the x-ray. If the uncertainty in the momentum of the atom is much greater than the momentum of the x-ray, then any attempt to measure the recoil is lost in the uncertainty principle.

The same effect occurs in kaon regeneration phenomena. Consider a beam of the long-lived neutral kaon state K_L passing through two plates of matter. The interaction of a K_L with matter can change it into the other short lived kaon state the K_S. When a beam passes through two plates, the two K_S waves produced in the two plates are coherent and can interfere. This results in an interference pattern which depends upon the distance between the two plates, because the K_L and K_S states have a small mass difference, and the phase of a wave passing between the two plates depends upon the mass.

But why are the two kaon waves coherent? Because the K_L and K_S have different masses, there must be a momentum transfer during the regeneration process, and a momentum measurement would determine which plate has produced the K_L-K_S transition. In this case the scattered kaon waves from the two plates would not be coherent.

Here again the uncertainty principle comes to the rescue. The plates are held in space with a very small uncertainty in position. This is reflected as a large momentum uncertainty. The key to all these phenomena is Eqs.(2). In any measurement of a wave length, the positions of the two plates must be known to much better than a wave length for the measurement to make any sense at all. The uncertainty principle then tells us that this localization of the plates produces an uncertainty in momentum much larger than any recoil momentum which would identify the plate responsible for the transition.

Another example of the "recoil-free" effect is in neutrino oscillations. It has been suggested that the two neutrinos associated with electrons and muons may not be both massless, but that at least one might have a finite mass, and that there may be a mass difference between them. Furthermore, the neutrino state produced in muon decay might not be a mass eigenstate but a coherent linear combination of the "heavy" and "light" neutrinos. Since the heavy and light states have different masses, like the K_S and K_L, they propagate with different phase, and "neutrino oscillations" might be observed. These oscillations would have a very long wave length, and there have been proposals to point the Fermilab neutrino beam at Argonne or somewhere in Canada in order to see whether the neutrino beam changes its properties over such large distances.

But why are the heavy and light neutrino beams coherent? If they are produced by pion decay into a muon and a neutrino, the momentum of the muon depends upon the mass of the neutrino, and a precise measurement of the muon momentum would determine the mass of the neutrino. In this case the waves of different masses would not be coherent.

But the uncertainty principle is here again! The initial pion which decayed to produce the neutrino beam must have its position well enough defined compared to the wave length that is being measured. Again Eqs. (2) apply. The position of a source where a transition is taking place must have an uncertainty much smaller than a characteristic wave length which is being measured. But the characteristic wave length is always complementary to the recoil momentum whose precise measurement would destroy the coherence.

We have heard a great deal about coherence and superradiance and about speeding up lifetimes and suppressing inelastic channels. To get a simple picture of these effects, let us look at the intensity of radiation scattered by an assembly of N coherent scatterers which are all in phase. The scattered intensity is obtained by summing the scattering amplitudes and squaring to get a term proportional to N^2,

$$I_{scatt} = |\sum_{i=1}^{N} a_{scatt}^i|^2 = N^2 |a_{scatt}|^2 \tag{3}$$

If there are absorption channels open at each scatterer, there is also an absorption amplitude. However, this amplitude refers to an inelastic channel, in which the final state of the scatterer is different from the initial state. The final state depends upon which scatterer has absorbed the photon. The absorption is thus an incoherent sum of the contribution from the N scatterers and is proportional to N.

$$I_{abs} = \sum_i |a_{abs}^i|^2 = N |a_{abs}|^2 . \tag{4a}$$

Thus

$$I_{abs}/I_{scatt} = \frac{1}{N}|a_{abs}/a_{scatt}|^2 \tag{4b}$$

The ratio of absorption to scattering goes as $1/N$ and is suppressed. But one sees immediately that something is wrong with this naive derivation of the suppression of inelastic channels. If the scattered intensity goes as N^2, what happens when $N^2 \to \infty$? The scattered intensity cannot be greater than the incident intensity. In Bragg scattering experiments one considers cases where 50% of the incident beam can be reflected in the Bragg peak. This intensity will not be multiplied by a factor of 4 if the number of scatterers is doubled.

Some insight into this paradox is obtained by looking at the time dependence of the scattered radiation. Consider the scattering amplitude as a function of time after a single scatterer is hit by a pulse of radiation very short in comparison with the lifetime of the scattering state (a Mössbauer resonance), but long compared to transit times of light through the crystal. Then we can write

$$a_{scatt}(t) = a(0) \, e^{-\Gamma t/2} \tag{5a}$$

where $a(0)$ denotes the amplitude at time $t = 0$, and the decay constant Γ is the width of the scattering level. If there are N scatterers, summing up the amplitudes and squaring gives

$$I_{scatt}(t) = N^2 |a(0)|^2 \, e^{-\Gamma t} \tag{5b}$$

This clearly goes wrong when N becomes too big. We shall see that the way it goes wrong is that Γ changes as a function of N. For times small compared to the lifetime, but large enough so that the photon has time enough to pass through the crystal, the transition probability per unit time is given by the famous Golden Rule formula.

$$W_{i_N \to f_N} = (2\pi/\hbar) |\langle f_N |M| i_N \rangle|^2 = (2\pi N^2/\hbar) |\langle f_N |m| i_N \rangle|^2 \tag{6}$$

where i_N and f_N denote the initial and final states at the N scatterer system and M is the transition operator for the system of N scatterers and m is the transition operator for a single scatterer. We thus see that the naive result (3) holds for times which are short enough compared to the lifetime before the exponential decay has become appreciable. But the decay rate and lifetime clearly depend upon the number of scatterers. How does this happen?

What is a lifetime? The picture of a nucleus being excited and remaining in the excited state for a period of time before decaying is misleading. One then gets the impression that a nucleus does not know how much time to wait before decaying because it depends upon how many other nuclei are around. To get a better picture of what a lifetime means in exponential decay, let us consider a very simple exponential decay, the discharge of a condenser.

If a condenser with capacity C has a charge Q_o on it at time $t = 0$ and then discharges through a resistance R, we know that the voltage initially on the condenser is

$$V_o = Q_o/C \tag{7a}$$

The current flowing out of the condenser is

$$I_o = (V_o/R) = (Q_o/RC) = dQ/dt \quad (t=0) \tag{7b}$$

If we know the current I_o at $t = 0$, we know how fast the condenser is initially discharging. But it does not continue to discharge at this rate, because the rate of discharge depends on how much charge is left in the condenser. The simple relation (7b) holds only near $t = 0$. But we can use it as a differential equation for Q at any t and derive the exponential decay.

The same principles of exponential decay apply to the "discharge" of a quantum system with an excitation. If the pulse of radiation is over at time $t = 0$, then we have the system in an excited state which emits radiation. The rate of emission of radiation is given by the Golden Rule formula, which is directly analogous to the relation (7a) for the rate of current flow out of the condenser. As transition probability flows out of the quantum system, the probability of it still remaining excited decreases. Thus we can write

$$W = (2\pi/\hbar)P|\langle f_N|M|i_N\rangle|^2 = -dP/dt \tag{8}$$

where P is the probability that the system is excited. If this is considered as a differential equation for P, we obtain the usual exponential decay formula, and we see that the time constant of the exponential decay is

$$\Gamma = (2\pi/\hbar)|\langle f_N|M|i_N\rangle|^2 = N^2 (2\pi/\hbar)|\langle f_N|m|i_N\rangle|^2 \tag{9a}$$

In order to see the dependence of Γ on N, we must express the result (9a) in terms of initial and final states of a single scatterer, i_1 and f_1.

$$\Gamma = N \ (2\pi/\hbar) \left| \langle f_1 | m | i_1 \rangle \right|^2 . \tag{9b}$$

The way in which normalization factors of $1/\sqrt{N}$ enter in obtaining this result (9b) is illustrated below in Eqs. (13) and (14) for the particular case of $N = 3$. Thus Γ does increase with N as we have conjectured above.

The resolution of the paradox is that no nucleus is waiting around a time interval to decay. It begins decaying immediately, with a probability given by the Golden Rule. The lifetime is simply the time required for the probability to flow out of the state, just as the lifetime of an RC circuit does not depend upon any inherent delay time, but on the time it takes for the current to flow out of the condenser at the rate given by the "electronics golden rule" called Ohm's Law. If there are more scatterers present, the probability for decay flows out faster, and it is "used up" in a shorter time.

It is always instructive in Mössbauer paradoxes to look at the phenomena from two different and complementary points of view. Let us now examine the speeding up of lifetimes and suppression of inelastic channels in the complementary energy domain. We examine the intensity of the scattered wave as a function of energy. With the same naive approach as that used to obtain Eqs. (5) - (9), we obtain

$$I_{scatt}(E) = \phi_{inc}(E) \ \Sigma \ (E) = \phi_{inc}(E) \ N^2 \ \sigma \ (E) \tag{10}$$

where $\phi_{inc}(E)$ is the incident flux at energy E per unit area, $\Sigma \ (E)$ is the coherent scattering cross section at energy E for the N-scatterer system all scattering in phase, and $\sigma \ (E)$ is the scattering cross section for a single scatterer. This expression is clearly wrong at resonance, where the cross section reaches the unitarity limit $\pi \ \lambda^2$, multiplied by appropriate angular momentum factors, independent of the composition of the scattering center,

$$\sigma \ (E_{res}) = \Sigma \ (E_{res}) = \pi \ \lambda^2 \tag{11}$$

Thus we see that when $\sigma \ (E)$ becomes so large that multiplying it by N^2 violates unitarity (the entire partial wave is already scattered and there is nothing left to be scattered), the simple relation (10) must break down. But far off resonance where the scattering is weak and far from the unitarity limit, the simple relation should hold. This is achieved, as in the time dependent formulation, by noting that the width of the resonance must be proportional to N. Thus if the resonance is isolated and described by a simple Breit-Wigner formula,

$$\sigma (E) = \pi \; \lambdabar^2 \; (\Gamma^2/4)/[(E - E_r)^2 + \Gamma^2/4] \qquad (12a)$$

Then the cross section for N scatterers is

$$\Sigma (E) = \pi \; \lambdabar \; \{(N\Gamma)^2/4\}/[(E - E_r)^2 + (N\Gamma)^2/4] \qquad (12b)$$

At resonance, the two cross sections are equal, but far off resonance the simple relation (10) holds.

Thus unitarity or conservation of probability broadens the resonance and speeds up the lifetime.

I conclude this talk with a look at superradiance in color. Let us consider three excitable Mössbauer nuclei which we identify by coloring them red, green and purple. The wave function for the coherent or "superradiant" state is

$$|\psi\rangle = \frac{1}{\sqrt{3}} \; [|R\rangle + |G\rangle + |P\rangle] \qquad (13)$$

where $|R\rangle$ is the state with the red nucleus excited, the others in the ground state, etc. Then the decay rate of this state into the ground state g is given by the Golden Rule,

$$W = \frac{2\pi}{\hbar} \; |\langle g|H_{int}|\psi\rangle|^2$$

$$= \frac{2\pi}{\hbar} \cdot \frac{1}{3} \; |\langle g|H_{int}|R\rangle + \langle g|H_{int}|G\rangle + \langle g|H_{int}|P\rangle|^2 \qquad (14a)$$

Since the interaction H_{int} does not depend upon the color of the nucleus,

$$W = \frac{2\pi}{\hbar} \cdot 3 \; |\langle g|H_{int}|R\rangle|^2 \qquad (14b)$$

We thus see that the decay rate is enhanced by a factor equal to the number of colors. As in Eqs. (9) a factor of N^2 comes from the matrix element, but there is also a factor of $1/N$ from the normalization of the coherent state (13).

Nor let us consider the decay $\pi^o \rightarrow 2\gamma$, where the π^o is a coherent state of

$|\pi_R^o\rangle$ made of red quarks,

$|\pi_G^o\rangle$ made of green quarks,

$|\pi_P^o\rangle$ made of purple quarks,

$$|\pi^o\rangle = \frac{1}{\sqrt{3}} \{|\pi_R^o\rangle + |\pi_G^o\rangle + |\pi_P^o\rangle\} \qquad (15)$$

Then the decay rate is

$$|\langle 2\gamma|H_{int}|\pi^{o}\rangle|^{2} = \frac{1}{3}\{\langle 2\gamma|H_{int}|\pi_{R}^{o}\rangle$$

$$+ \langle 2\gamma|H_{int}|\pi_{G}^{o}\rangle + \langle 2\gamma|H_{int}|\pi_{P}^{o}\rangle\}^{2} \tag{16a}$$

If H_{int} does not depend upon color,

$$|\langle 2\gamma|H_{int}|\pi^{o}\rangle|^{2} = 3 \ |\langle 2\gamma|H_{int}|\pi_{R}^{o}\rangle|^{2} \tag{16b}$$

Similarly for the $\pi^{+} \to \mu^{+}\nu$ decay,

$$|\langle \mu^{+}\nu|H_{int}|\pi^{+}\rangle|^{2} = 3 \ |\langle \mu^{+}\nu|H_{int}|\pi_{R}^{+}\rangle|^{2} \tag{17}$$

A theoretical calculation of these decays turns out to be really a calculation of the product $\Gamma(\pi^{o} \to 2\gamma) \times \Gamma(\pi^{+} \to \mu\nu)$. This is enhanced by factor 9 if there are three colors of quarks! Experimental results are in striking agreement with the three color model.

CAN THE SIGN OF THE EFG BE OBTAINED FROM

COMBINING TDPAC AND MÖSSBAUER TECHNIQUES ?

G.R. Hoy[*] and J. Chappert

DRF, Groupe Interactions Hyperfines

Centre d'Etudes Nucléaires, 85 X-38041 Grenoble, France

H.C. Benski

Université Scientifique et Médicale, Grenoble, France

ABSTRACT

We have studied a proposed new technique for measuring the sign of the electric field gradient (EFG) by combining the methods of time differential perturbed angular correlation (TDPAC) and Mössbauer spectroscopy. Our experiments do not confirm such a suggestion and our recent calculations show that the effect does not exist.

INTRODUCTION

The idea of combining angular correlation with Mössbauer spectroscopy has existed for many years[1]. Recently the proposal was made[2] to use this method to determine the sign of the EFG in polycrystalline samples. According to reference 2 using a quadrupole split polycrystalline source of, for example, ^{57}Co and a single line absorber, one would expect the resulting two lines of the ^{57}Fe doublet spectrum to have relative intensities which are a function of the angle between the two detectors. Two of us (J.C. and H.C.B.) have done such experiments using sources of ^{57}Co in beryllium and cobalt phthalocyanine. The asymmetry in the intensities of the two lines was never statistically significant and did not allow any definite conclusions regarding the sign of the EFG.

* Permanent address: Boston University (USA). Supported in part by the National Science Foundation Grant No. DMR 73-07665A03.

RECENT RESULTS

A second experimental approach consisting of observing the lifetime curve, for example, of the 14.4 keV level in ^{57}Fe, through a Mössbauer resonant filter was also recently proposed[3]. This approach was also tested, using the experimental set-up shown in Fig. 1. The lifetime curve is recorded through a nuclear resonant filter (enriched sodium ferrocyanide) moving at constant velocity to match one of the two transitions in the source (^{57}Co in beryllium). Time filtering effects[4] drastically affect the shape of the "lifetime curve". Our experimental results are shown in Fig. 2. Although some small differences appear in both geometries (90° and 180°) between the "lifetime curves" recorded for each component, they were not apparently due to the quadrupole interaction. One of us (G.R.H.) made a calculation of the proposed effect assuming $\eta = 0$ and a pure M1-M1 cascade and at first obtained the same results as in reference 3. However, (see acknowledgements), an incorrect time dependence was found for the angular correlation of the 122 keV–14.4 keV γ-γ cascade through a particular Mössbauer quadrupole slit level (say $\pm$ 3/2) in the intermediate state. This observation lead to the detection of a mistake in our calculation. The basic point is that in Mössbauer TDPAC effect, the Mössbauer effect measures an energy. We are assuming here that the actual hyperfine interaction produces a splitting of some intermediate state levels larger than the line width i.e. different transitions can be distinguished by the Mössbauer effect. This means that in the calculation of the result through, for example, the $\pm$ 3/2 "channel" it is only those "m" values which are associated with states that are diagonal with respect to the hyperfine interaction that take on the values $\pm$ 3/2. This has the effect of removing the above mentioned time dependence. In addition any angular dependence found in the $\pm$ 3/2 channel is the same as in the $\pm$ 1/2 channel. Thus the angular dependence with respect to the direction of the detected 122 keV photons is the same for the 14.4 keV radiation coming from the $\pm$ 3/2 level or the $\pm$ 1/2 level. In that case the sign of the EFG can not be determined.

The mathematical origin of the equality of the two angular distributions is due to the following identity of 3 j symbols;

$$
\begin{pmatrix} 3/2 & 3/2 & k \\ 3/2 & 3/2 & -3 \end{pmatrix}^2 + \begin{pmatrix} 3/2 & 3/2 & k \\ -3/2 & -3/2 & 3 \end{pmatrix}^2 + \begin{pmatrix} 3/2 & 3/2 & k \\ 3/2 & -3/2 & 0 \end{pmatrix}^2 + \begin{pmatrix} 3/2 & 3/2 & k \\ -3/2 & 3/2 & 0 \end{pmatrix}^2 =
$$

$$
\begin{pmatrix} 3/2 & 3/2 & k \\ 1/2 & 1/2 & -1 \end{pmatrix}^2 + \begin{pmatrix} 3/2 & 3/2 & k \\ -1/2 & -1/2 & 1 \end{pmatrix}^2 + \begin{pmatrix} 3/2 & 3/2 & k \\ 1/2 & -1/2 & 0 \end{pmatrix}^2 + \begin{pmatrix} 3/2 & 3/2 & k \\ -1/2 & 1/2 & 0 \end{pmatrix}^2
$$

for each value of k = 0, 1, 2, 3. This mathematical result was not known to us, but just appeared as a sufficient condition in the calculation.

ACKNOWLEDGEMENT

One of us (G.R.H.) would like to acknowledge important discussions with R. Coussement and G. Langouche who were very skeptical of the time dependences in the original calculation. We also thank A.M. Afanasev for a helpful remark. Technical assistance by J. Berthier and Y. Gros is gratefully acknowledged. J.A. Sawicki was kind enough to supply the [57]Co beryllium source.

REFERENCES

1. See for example, G.R. Hoy, D.W. Hamill and P.P. Wintersteiner "Mössbauer Effect Methodology", Vol. 6, edited by L.G. Gruverman (Plenum, New York 1971).

2. A.Z. Hrynkeiwicz, E. Popiel and J. Sawicki, Acta. Phys. Polonica A 45, 307 (1974).

3. E. Popiel and J.A. Sawicki, Intern. Conf. Mössbauer Spectroscopy, Cracow, Vol. I, p. 13 (1974).

4. F.J. Lynch, R.E. Holland and M. Hamermech, Phys. Rev. 120, 513 (1960).

Fig.1 A schematic diagram of our experimental set-up for measuring a lifetime curve through a nuclear resonant filter moving at constant velocity. The gating pulse is used to record only those events corresponding to the filter moving at the correct constant velocity.

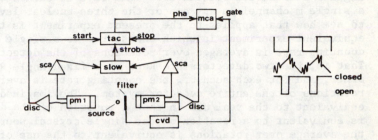

Fig. 2. Our experimental results are shown here. The solid points correspond to the case when the filter velocity matches the lower energy component of the doublet spectrum. The open circles show the result when the filter is "set on" the higher energy component. The enriched sodium ferrocyanide filter has an effective thickness $T_a \sim 14$.

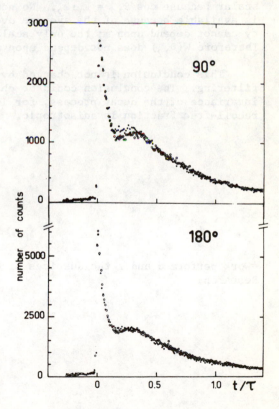

COMMENT TO THE PAPER OF HOY, CHAPPERT, AND BENSKI

Murray Peshkin
Argonne National Laboratory, Argonne, Illinois 60439*

It is a general result of the Wigner-Eckart theorem that the angular correlation between two gamma rays is unchanged by selecting a single $\underline{m}$ channel in any one of the three nuclear levels. One way to see how that applies to the present experiment is to imagine an equivalent experiment in which the source is a single crystal but the counting rate is averaged over rotations of the detector system. That is, the two detectors are kept at a fixed angle θ_{12} from each other, and for each such θ_{12} the counting rate is averaged over rigid rotations of the entire detector system. The imagined experiment is equivalent to the real one in that a rotation of the detector system is equivalent to a rotation of the single crystal source, so that the average over rotations is equivalent to the use of a powdered crystal source.

In the imagined experiment, there is a single $\underline{z}$ axis and the meaning of $\underline{m}$ is clear. The angular correlation function $W(\theta_{12})$ is scalar because $\cos\theta_{12} = \hat{k}_1 \cdot \hat{k}_2$. No non-scalar angular information is available because of the average over rotations. A scalar quantity cannot depend upon $\underline{m}$; the only scalar function of $\underline{m}$ is a constant. Therefore $W(\theta_{12})$ does not depend upon $\underline{m}$.

This conclusion is not changed by multipole mixing or by time filtering. The conclusion could be changed by spoiling the rotation invariance of the decay process, for instance in the case where the recoil-free fraction is anisotropic.

*Work performed under the auspices of USERDA, Div. of Physical Research.

Dispersion Amplitudes in the Mössbauer Lineshape of ^{237}Np, ^{182}W and ^{195}Pt

W. Potzel, F.E. Wagner, G.M. Kalvius

Physik Department, Technische Universität München,
D-8046 Garching, Germany

L. Asch

Laboratoire de Chimie Nucléaire, F-67037 Strasbourg,
France

J.C. Spirlet, W. Müller

European Institute for Transuranium Elements,
D-7500 Karlsruhe, Germany

Abstract

We report on measurements of the dispersion amplitude
for the Mössbauer resonances in ^{237}Np (60 keV), ^{182}W
(100 keV) and ^{195}Pt (99 keV). For the E1 transition in
^{237}Np a value of $\xi = -(3.4 + 0.2) \times 10^{-2}$ was found.
Theoretical estimates of ξ for this case are of limited
accuracy and the comparison with experimental data is
inaccurate because of the considerable line broadening.
Computer simulations show the influence of ξ on ^{237}Np hyperfine spectra. For the E2 transition in ^{182}W the dependence of the effective value of ξ on absorber thickness
was studied. Extrapolations to zero thickness yield
$\xi = -(1.0 \pm 0.1) \times 10^{-2}$ in good agreement with theoretical calculations. For ^{195}Pt we found $\xi = -(1.1 \pm 0.3)$
$\times 10^{-2}$ for the dispersion amplitude at zero absorber
thickness. This is the first pure M1 transition for
which a dispersion term has been observed. For all three
resonances the dispersion is large enough to affect the
determination of isomer shifts.

Introduction

The existence of a dispersion term in addition to the Lorentzian energy dependence of the cross section for nuclear resonance absorption has been established both theoretically[1-4] and experimentally[5-12]. This phenomenon can be visualized as arising mainly from the interference between nonresonant emission of a photoelectron and resonant excitation of the nucleus followed by an internal conversion process. The absorption cross section for a single, isolated resonance then takes the form

$$\sigma_a = \sigma_0 \frac{1 + 2\xi x}{x^2 + 1}$$

where σ_0 is the cross section at resonance and $x = 2(v-S)/W$, S being the isomer shift, W the full resonance width at half maximum in velocity units, and v the Doppler velocity.

The dispersion amplitude ξ is most prominent for Mössbauer transitions with E1 multipolarity since the photoelectric process is largely of electric dipole character. It is smaller, but still noticeable, for E2 and mixed M1+E2 transitions. Neglecting the dispersion term in the analysis of hyperfine spectra may lead to erroneous results, particularly with respect to the isomer shift. It may affect other hyperfine parameters as well whenever the spectra are poorly resolved.

Of the well established E1 Mössbauer resonances, only the 60 keV transition in ^{237}Np has not yet been studied with special emphasis on the dispersion effect. We now present extensive measurements of the dispersion amplitude for this case. The dispersion term for the 100 keV E 2 transition in ^{182}W has been investigated previously[8]. For this case, we have studied the change of the effective dispersion amplitude with

absorber thickness. This problem has been dealt with theoretically by various authors[13,14], and experimental data
are available for some resonances of E1 multipolarity
(e.g. ^{181}Ta[12] and ^{155}Gd[6]). The value of ξ extrapolated to
zero absorber thickness can be compared with recent calculations[4].

For M1 transitions no dispersion terms have been observed
sofar, but calculations[4] predict that transitions of relatively high energy in heavy nuclei should exhibit measurable interference effects. Our experiments on the 99 keV
M1 transition in ^{195}Pt confirms this prediction. Since
this resonance has often been used in investigations of
isomer shifts (see, e.g., ref.[15]), the knowledge of the
dispersion term is of some significance.

Experimental

a) $\underline{^{237}Np}$: Two sources of Am metal were used. One of them
consisted of 300 mC of ^{241}Am evaporated onto a Ta disc of
11 mm diameter, the other of 50 mC of ^{241}Am evaporated onto a Ta disc with only 4 mm diameter. The sources were
kept at either 4.2 K or 77 K. As the resonance absorber
we used 100 mg/cm^2 of $^{237}NpO_2$ pressed into a disc of 25mm
diameter. Since NpO_2 is known to become magnetic near 20 K,
the absorber was always kept at 77 K. Special care was ex-
ercised to ensure a good beam geometry in order to obtain
a straight off-resonance baseline. The sinusoidal Doppler
motion[16] was fed to the source. Its frequency was chosen
rather high (70 Hz) to keep the amplitude of motion, and
with it the geometry effect, as small as possible. The
gamma rays were counted with a 5 % tin loaded plastic
scintillator in connection with a fast level-discrimina-
tor[17]. In this way countrates up to 10^6 s^{-1} could be handled without lineshape distortions. The spectra were recorded in a PDP8e computer operated in time mode[18].

b) ^{182}W: The ^{182}W Mössbauer spectra were taken with single-line absorbers of natural tungsten metal and a single line source of ^{182}Ta produced by neutron activation of a Ta metal foil. In these experiments the source and absorber were always kept at 4.2 K.

c) ^{195}Pt: For the experiments with the 99 keV transition in ^{195}Pt, sources of ^{195m}Pt ($T_{1/2}$ = 4 d) were used. These were prepared by neutron activation of Pt metal enriched in ^{194}Pt to 97.4 %. The absorbers were natural Pt metal foils. Sources and absorbers were cooled to 4.2 K.

Results and Discussion

a) ^{237}Np: In Fig. 1 we show fits of a typical spectrum with and without a dispersion term added to a single Lorentzian. The improvement obtained by including the dispersion term is clearly visible. A summary of the experimental data can be found in table 1. Each of the three entries is the average of at least four independent runs.

Table 1: Dispersion amplitudes for ^{137}Np

Source	Source Temp.	Counts per point	ξ
300 mC	4.2 K	3×10^8	$-(3.3 \pm 0.2) \times 10^{-2}$
300 mC	77 K	5×10^8	$-(3.4 \pm 0.3) \times 10^{-2}$
50 mC	77 K	4×10^8	$-(3.7 \pm 0.4) \times 10^{-2}$
		Average: ξ_{ex}	$= -(3.4 \pm 0.2) \times 10^{-2}$

The final result for ξ_{ex} can be compared with theoretical estimates. In doing so one should, however, keep in mind that the experimental value for the dispersion parameter has been obtained for a line that is 20 to 30 times broader than the natural width. The reasons for this broadening are not understood in detail, and no way of obtaining narrower lines is presently known for the ^{237}Np resonance. In princip line broadenings do not seriously affect the magnitude of

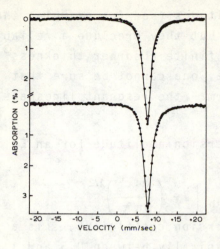

Fig. 1: Mössbauer spectrum measured with the 300 mC ^{241}Am source at 4.2 K and a 100 mg/cm^2 ^{237}NpO$_2$ absorber at 77 K. The total number of counts per velocity point is 3×10^7.
Curve (a) shows a fit to the data points without dispersion, yielding W = 2.34 mm/s and S = 7.93 mm/s. Curve (b) is a fit to the same data including dispersion. The fit parameters are W = 2.33 mm/s, S = 7.85 mm/s and ξ = - 0.033.

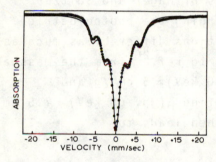

Fig. 2: Simulated electric quadrupole hyperfine spectrum including dispersion (dots). The input parameters are given in line 2 of table 2. The curve is a fit of a purely Lorentzian lineshape to the simulated data.

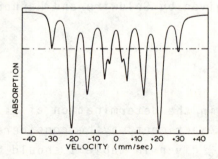

Fig. 3: Simulated magnetic hyperfine pattern for the ^{237}Np resonance. The input parameters for this spectrum have been B$_{eff}$ = 100 T, W = 2.3 mm/s, S = 0 mm/s and ξ = - 0.03.

the measured dispersion amplitudes (see, for example, the cases of ^{181}Ta[10] or ^{155}Gd[6]), but they preclude a reliable correction for the effect of finite absorber thickness[13,14] (see also below). Furthermore, one cannot be sure that - except for the dispersion term - the resonant lineshape is exactly Lorentzian.

According to ref.[3], the dispersion amplitude for an E1 transition can be expressed as

$$\left| \xi_{th} \right| = \epsilon \left(\frac{\alpha \cdot \sigma'_e}{6 \pi \lambdabar} \right)^{1/2}$$

where α is the internal conversion coefficient, ϵ is a statistical factor that is typically between 0.5 and 1, and σ_e' is the partial cross section for E1 photoelectric absorption. We will assume that the photoelectric process in Np is of pure E1 character (i.e. $\sigma_e' = \sigma_{ph}$). Furthermore we take α from the tables of Hager and Seltzer[19]. A large uncertainty exists regarding the photoelectric cross section. Various tabulations differ by as much as 40 %. Extrapolating according to ref.[20] from the mass absorption coefficients $\mu(Pb, 60 \text{ keV}) \approx 5 \text{ cm}^2/g$ and $\mu(U, 60 \text{ keV}) \approx 6.6 \text{ cm}^2/g$[21] we find $\mu(Np, 60 \text{ keV}) \approx 7.5 \text{ cm}^2/g$ and hence $\sigma_{ph} \approx 3000$ b. This then leads to

$$\left| \xi_{th} \right| \leq 0.05$$

One can also estimate ξ_{th} for Np by extrapolation and interpolation from the tables given by Goldwire and Hannon[4]. This yields

$$\xi_{th} = -3 \times 10^{-2}$$

In view of the uncertainties in the determination of ξ, th agreement between theory and experiment may be considered satisfactory. The value of ξ_{ex} given in table 1 should be taken into account in the analysis of Np hyperfine spectra

The dominant effect is clearly in the determination of isomer shifts. For example, the spectrum of Fig. 1 yields S = 7.93 mm/s for a pure Lorentzian fit and S = 7.86 mm/s when the dispersion term is included. In Fig. 2 we have simulated (data points) a pure quadrupole spectrum including dispersion, which was then fitted with a sum of Lorentzian lines (full line). In table 2 we have summarized the results of such simulations. One again sees that mainly the isomer shift is returned erroneously. The line intensity ratios for the individual peaks of poorly resolved hyperfine patterns are also seriously affected. We illustrate the effect on line intensities by a different example. Fig. 3 shows a fairly well-resolved magnetic hyperfine pattern as expected for B_{hf} = 100 T. Assuming Lorentzian lineshapes, one might interpret the difference in the depth of the outermost resonance lines as arising from nuclear polarization[22]. The deduced intensity ratio, I_-/I_+ = 0.895, would - in the absence of dispersion - be expected for a sample temperature of 1.85 K.

Table 2: Simulations of Quadrupole Spectra

Input Parameters for Simulation				Parameters returned by Lorentzian fit			
e^2qQ	S	W	ξ	e^2qQ	S	W	ξ[1)
mm/s	mm/s	mm/s		mm/s	mm/s	mm/s	
30	0	2.3	-0.05	30.0	0.11	2.29	0
20	0	2.3	-0.05	19.9	0.13	2.30	0
15	0	2.3	-0.05	14.8	0.16	2.27	0
8	0	2.3	-0.05	7.9	0.19	2.31	0

[1)]constrained to zero

b) [182]W: The values of ξ measured for absorbers of different thickness are shown in Fig. 4. These data enable us to extrapolate to zero absorber thickness. This yields

ξ (t=0) = - (1.0 $\pm$ 0.1) x 10^{-2}, in good agreement with the theoretical value, ξ_{th} = - 1.2 x 10^{-2} calculated by Goldwire and Hannon[4] for this E2 transition. As has been shown both empirically[8] and by computer simulations[23], the apparent isomer shift resulting when the dispersion is neglected in least-squares fits of the spectra is well approximated by S = ξ · W. With W $\approx$ 2.5 mm/s for the ^{182}W resonance, these shifts are comparable to the true isomer shifts observed in this case[24].

c) $\underline{^{195}Pt}$: For the 99 keV resonance in ^{195}Pt measurements with absorbers containing 200 and 430 mg/cm^2 of natural

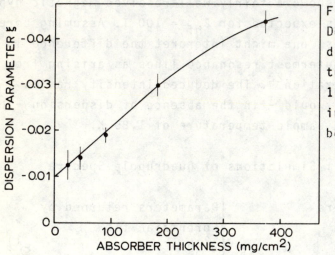

Fig. 4:
Dependence of the dispersion amplitude for the 100 keV transition in ^{182}W on absorber thickness.

Pt metal yielded ξ = - (1.5 $\pm$ 0.5) x 10^{-2} and - (2.5 $\pm$ 0.5) x 10^{-2}, respectively. These values are not sufficiently accurate for an empirical extrapolation to zero absorber thickness, but since the resonance line is not broadened by hyperfine interactions, the correction for finite absorber thickness can be made with the relation

$$\xi(t) = \xi(t=0)(1+0.27t)$$

obtained[13] from numerical lineshape calculations. Here
$t = f \cdot n \cdot \sigma_o$ is the effective thickness of an absorber containing n resonant nuclei per unit area. Using $f = 0.136$
for the f-factor of Pt metal[15], we obtain $t = 1.7$ and
$t = 3.7$ for our absorbers, and hence $\xi(t=0) = -(1.0\pm0.4)$
x 10^{-2} and $-(1.2 \pm 0.3)$ x 10^{-2}, respectively. The final
average result, $\xi(0) = -(1.1 \pm 0.3)$ x 10^{-2}, is again in
fair agreement with the value of $\xi_{th} = -0.7$ x 10^{-2} that
can be estimated by extrapolation and interpolation from
the calculations of Goldwire and Hannon[4]. The 99 keV transition in ^{195}Pt is actually the first pure M1 transition
for which a dispersion term has been observed. With experimental linewidths of $W \approx 20$ mm/s, this dispersion term
may bring about apparent isomer shifts of the order of a
few tenths of a mm/s. These may, at least in certain
cases, be of importance in isomer shift studies[15].

References

1. G.T. Trammel and J.P. Hannon, Phys. Rev. Letters 21,
   726 (1968); Phys. Rev. 180, 337 (1969)

2. Yu. Kagan, A.M. Afanas'ev and V.K. Voitovetskii
   Soviet Physics: JETP-Letters 9, 91 (1969)

3. J.P. Hannon and G.T. Trammel, Phys. Rev.186,306(1969)

4. H.C. Goldwire jr. and J.P. Hannon, to be published

5. C. Sauer, E. Mathias and R.L. Mößbauer, Phys. Rev.
   Letters 21, 961 (1968)

6. W. Henning, G. Bähre and P. Kienle, Phys. Letters
   31B, 203 (1970)

7. D.V. Gorobchenko, I.I. Lukashevich, V.V. Sklyarevskii
   and N.I. Filippov, Soviet Physics: JETP Letters 9,
   139 (1969)

8. F.E. Wagner, B.D. Dunlap, G.M. Kalvius, H. Schaller,
   R.Felscher and H.Spieler, Phys.Rev.Letters 28,530(1972)

86

9. D.J. Erickson, J.F. Prince, and D.L. Roberts, Phys. Rev. C8, 1916 (1973)

10. G. Kaindl, D. Salomon and G. Wortmann in:"Mössbauer Effect Methodology", Vol.8 (I.J. Gruverman and C.W. Seidel, eds.) Plenum Press, New York 1973

11. L. Pfeiffer, Phys. Rev. Letters 38, 862 (1977)

12. D. Salomon, P.J. West and G. Weyer, Hyperfine Interactions, to be published

13. P. West, Nucl. Instr. Meth. 101, 243 (1972)

14. B.T. Cleveland and J. Heberle, Physics Letters 40A, 13 (1972)

15. D. Walcher, Z. Phys. 246, 123 (1971)

16. N. Halder and G.M. Kalvius, Nucl. Instr. Meth. 108, 161 (1973)

17. G.M. Kalvius, W. Potzel, W. Koch, A. Forster, L. Asch, F.E. Wagner and N. Halder, contribution to this conference

18. A. Forster, N. Halder, G.M. Kalvius, W. Potzel and L. Asch, Jour. Physique 37, C6-725 (1976)

19. R.S. Hager and E.C. Seltzer, Nuclear Data A4, No 1-2 (1968)

20. M. Davisson in K. Siegbahn "Alpha-, Beta- and Gamma-Ray Spectroscopy", Ch. II. North Holland, Amsterdam 1965

21. Table of Mass Absorption Coefficients, Norelco Reporter, May-June 1962

22. G.M. Kalvius, T.E. Katila and O.V. Lounasmaa in: "Mössbauer Effect Methodology", Vol.5 (I.J. Gruverman ed.) Plenum Press, New York 1969

23. M. Karger, thesis, Technical University of Munich,1975 (unpublished)

24. F.E. Wagner, M. Karger, M. Seiderer and G. Wortmann, contribution to this conference

MÖSSBAUER EXPERIMENTS IN ANALOGY TO OPTICAL ACTIVITY

U. Gonser, H. Engelmann,
Universität des Saarlandes, 6600 Saarbrücken, Germany

H. Brunner, M. Muschiol, W. Nowak
Universität Regensburg, 8400 Regensburg, Germany

INTRODUCTION

The dispersion associated with Mössbauer resonance absorption has been measured by birefringence rotation polarimetry, particularly, by the Mössbauer Faraday effect [1-3]. It seems of interest to search for other effects in analogy to optics with resonant γ-rays of 0.86 Å wave length resulting from the 14.4 keV excited state of Fe^{57}. Specifically, experiments were carried out to observe Mössbauer "optical" activity by using optically active chiral molecules containing Fe^{57}. From classical theory of electromagnetic radiation one might expect an optical rotatory dispersion curve ($n_R - n_L$) around the resonance wave length, λ_o, reflecting the difference of the dispersion curves of left- and right- circularly polarized radiation as schematically indicated in fig. 1.

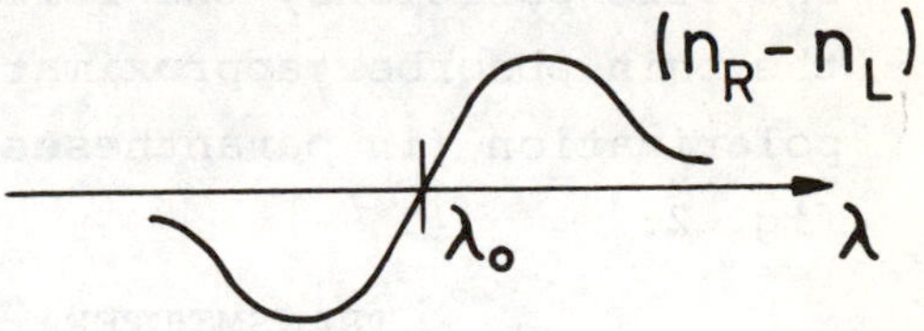

Fig. 1

It should be pointed out that the linearly polarized γ-radiation from the source (polarizer) can be represented as a superposition of right and left circularly polarized γ-radiation, and the optical effects of the γ-radiation transversing the transmitter can be understood by considering the corresponding indices of refraction.

MÖSSBAUER POLARIMETER

Co^{57}-α-Fe and Fe^{57} enriched α-Fe foils are
used as source (polarizer)
and absorber (analyzer),
respectively. Both are
transversely magnetized
by the fields H_S and H_A.
The propagation direction
of the γ-radiation (γ-
arrow, $\vec{\gamma}$) is perpendicular to H_S and H_A. The po-
larimeter is based on a
source emitting linearly
polarized recoil-free γ-
rays and a rotating ab-
sorber. In the fully automatic measuring procedure[4]
the counts are accumulated

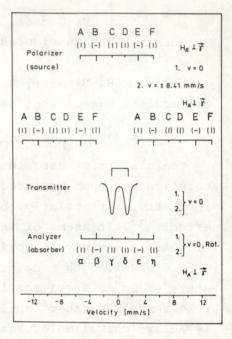

Fig. 2

in a multichannel analyzer with each channel corresponding to a well-defined angle, φ, between H_S in the
source (polarizer) and H_A in the absorber (analyzer).
The line positions, the relative line intensities - in
the thin absorber approximation - and the orientation of
polarization (in parantheses) are schematically shown in
fig. 2.

TRANSMITTER

The transmitter consists of the optically active iron
compound $(-)_{365}$-$C_5H_5Fe(CO)(I)P(C_6H_5)_2N(CH_3)CH(CH_3)(C_6H_5)$
enriched in Fe^{57}. In the preparation the two resulting
$(+)_{365}$- and $(-)_{365}$-diastereoisomers were separated by
fractional crystallisation.

EXPERIMENTS AND RESULTS

Two experiments were carried out under conditions schematically indicated in fig. 2.

1. Source, transmitter and absorber do not move in the direction of $\vec{\gamma}$ (v=o). The linearly polarized γ-rays, emitted by the source, are absorbed by the rotating absorber. The γ-rays corresponding to the weak lines, C, D are absorbed to a great extend by the transmitter, while the γ-rays corresponding to the main lines (A, B, E and F) are passing the transmitter on both sides of the spectrum and are in resonance with the absorber lines α, β, ε and η. Maximum resonance absorption occurs when the linearly polarized components are matched ($H_S \parallel H_A$). The relative transmission as a function of rotation, ω, of H_S compared to H_A is shown in fig. 3(a). The resulting "Malus curve" (dashed line) is fitted by a least square sinusoidal fit procedure and compared to the Malus curve without transmitter. A shift indicating optical activity could not be detected within our accuracy of about $\pm 1°$.

2. Transmitter and absorber do not move in the direction of $\vec{\gamma}$ (v=o), and, the absorber is rotating again. However, the source vibrates ($\approx 2o$ Hz) and moves with the two constant velocities (v=$\pm$8.41 mm/sec) to bring into resonance the A, B or E, F lines of the source with the ε, η or α, β lines of the absorber, respectively. In this case maximum resonance absorption of the linear polarization of source and absorber lines (Aε, Bη, Eα, Fβ) occurs when the orientation of H_S and H_A are perpendicular to each other ($H_S \perp H_A$). The emitted γ-rays with v=+8.41 mm/sec where Aε and Bη are in resonance (right side of fig. 2) are accumulated in one half of a multichannel analyzer and correspondingly, the γ-rays emitted with v=-8.41 mm/sec where Eα, Fβ are in resonance (left side of fig. 2) in the other half. The resulting Malus

90

curves (fig. 3(b)) in the two halfs of the multichannel analyzer represent the γ-radiation having passed the transmitter on the two sides of positive (top) and negative (bottom) velocity. This procedure has the advantage that any misalignment of source, transmitter and absorber (geometry effects) influences the amplitudes of both Malus curves in the same fashion. By comparing the two curves

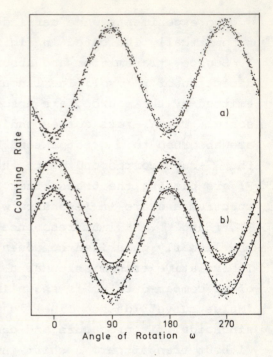

Fig. 3

in regard to their respective maxima and minima any shift indicates the effect of Mössbauer optical activity. In our experiment the fitting of the two Malus curves indicated a shift of 2.1 $^\circ$, however, ths result is within the experimental error of $\pm$ 1.5 $^\circ$ for each curve.

REFERENCES

1. P. Imbert, J. Phys. <u>27</u>, 429 (1966)
2. R. M. Housley, U. Gonser, Phys. Rev. <u>171</u>, 48o (1968)
3. M. Blume, O. C. Kistner, Phys. Rev. <u>171</u>, 417 (1968)
4. H. D. Pfannes, U. Gonser, Nucl. Instr. Methods <u>114</u>, 297 (1974)

The support by the DFG (SFB 130) is gratefully acknowledged.

PRELIMINARY REPORT ON AN EXPERIMENT TO EXCITE NUCLEAR RESONANCE BY SYNCHROTRON RADIATION

R. L. COHEN,   G. L. MILLER,   K. W. WEST

Bell Laboratories, Murray Hill, N. J., 07974*

A second run to excite nuclear levels via synchrotron radiation was carried out in May 1977. These experiments used the configuration shown in Fig. 1 to excite the 14 keV 100 nsec state of Fe^{57}, and observe the subsequent decay of this state via the emission of conversion electrons.

The background of photoelectrons (prompt with the X-ray pulse) is about $10^6\times$ larger than the signal from the conversion electrons. To discriminate against this background, both the electron multiplier and counting electronics are gated off during the arrival time of the X-ray pulse, but turned on within 100 nsec to observe the conversion electrons emitted in the decay of the 14 keV state as indicated in the Fig. 2 timing diagram.

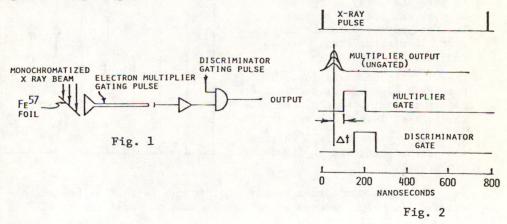

Fig. 1

Fig. 2

This double gating system is necessary to reduce the afterpulsing rate of the electron multiplier, which must be on the order of 10^{-5} or less to observe the delayed conversion electrons. The initial experiments (curve A, Fig. 3) showed a relatively high afterpulsing rate, and the multiplier gating was revised. Curve B shows the performance of the revised detector with a flux of 1.5×10^9 14 keV X-rays/sec incident on the foil.

* Partially supported by NSF Grant DMR73-07692 in cooperation with SLAC and U. S. ERDA.

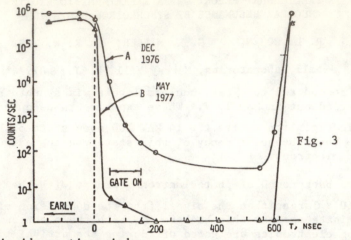

Fig. 3

With the coincidence time window set as shown in the figure, the monochromator energy was scanned to make the X-ray energy pass through the nuclear resonance. The data obtained in the first scan, 15 minutes long, were as indicated in Fig. 4.

The electron beam was lost, and the SPEAR high energy terminated before another scan could be made. The peak at 14.413 keV was at the known energy of the Fe^{57} resonance, and of the predicted amplitude of 0.6 c/sec.

$E_\gamma = 14.41303$

Fig. 4

The most significant result of this run is that we have demonstrated that the "between pulse" background of radiation at SSRP is at least 10^5 to 10^6 less than the pulse intensity. This value, which is important for certain fluorescence experiments, was not previously established. In reducing the rate of unwanted counts between the firs run (Fig. 3, curve A) and the second run (Fig. 3, curve B) we have found that there is a significant background arising from low energy ions or electrons produced by the X-rays. We will be investigating this phenomenon off-line before the next run at SSRP.

With the performance obtained in the May 1977 run, the detector could be useful in establishing the profile and energy calibration of the monochromator. The detection efficiency can readily be improved b a factor of two by the use of a larger effective foil area. Possible reduction of the background counting rate could also provide great increases in effective sensitivity, and provide a new approach for studies of energy dispersive monochromators.

FAST DATA ACQUISITION IN MÖSSBAUER SPECTROSCOPY

G.M. Kalvius[+], W. Potzel[+], W. Koch[+], A. Forster[+],
L. Asch[+*], F.E. Wagner[+] and N. Halder[+]
[+] Physik Department, Technische Universität München,
D-8046 Garching, Germany
[*] Laboratoire de Chimie Nucléaire, F-67037 Strasbourg,
France

ABSTRACT

Three different gamma-ray detection systems for Mössbauer
measurements at count rates above 10^5 s^{-1} are described. The
first uses the fast output of an uncooled NaI(Tl) scintillation
counter and will still provide energy discrimination at total
pulse rates up to 10^6 s^{-1}. Count rates up to some 10^7 s^{-1} can
be handled by a tin-loaded plastic scintillation detector, but
only with very limited energy resolution. For even higher
photon intensities, the method of current integration is available. The various techniques are illustrated and compared on
the basis of measurements with the Mössbauer resonances in
^{170}Yb (84 keV), ^{199}Hg (158 keV) and ^{237}Np (60 keV).

INTRODUCTION

The usual counting procedures in Mössbauer spectroscopy
make use of gamma-ray spectrometers with good energy resolution
but are restricted to total count rates of the order of 10^5 s^{-1}.
The data storage system can usually handle count rates between
10^6 to 10^7 s^{-1} and is thus no serious limitation to an increase
of counting speed. The measurements of small absorption effects,
or the accurate determination of spectral shapes, often requires
the collection of a large number of counts. Actually, the time
necessary to obtain a spectrum with a given accuracy is one of
the most serious limitations of Mössbauer spectroscopy. It
prevents, for example, the study of reactions which take place
on a time scale of seconds or minutes. The rather long times
needed for data acquisition have led to the development of
highly stable Doppler spectrometers (e.g. [1,2,3]), but counting
times in excess of a few weeks are usually impracticable.

As a measure of quality of the spectrum one may define [3]
the signal-to-noise ratio $S/B = \epsilon\sqrt{R \cdot T/2}$. Here R is the count
rate per multiscaler channel, T the total time spent for data
acquisition and $\epsilon = (N_\infty - N_0)/N_\infty$, where N_∞ and N_0 are the total number of
counts collected far off and exactly at resonance, respectively. In
order to measure an effect of 5×10^{-4} with an accuracy of $(S/B) = 4$ one
week is needed, if one collects $\sim 10^5$ counts/s in 512 channels.

Recently, several ways to improve the counting speed have
been suggested. NaI(Tl) scintillators can be made faster by
cooling [4], while good energy resolution is maintained. Plastic
scintillators have an inherent shorter decay time for their
fluorescent light and are thus faster. Their extremely poor

efficiency for gamma-rays and their nearly complete lack of
energy resolution can be overcome at energies below 200 keV
by loading the plastic with tin or lead. Still the energy
resolution remains by far inferior compared to NaI(Tl) scintillators.[5] Applications to Mössbauer spectroscopy have
only been reported [6] for ^{57}Fe and ^{119}Sn, which are poor
examples because the source strength is limited. Finally,
with the current integration technique that has been described in detail by Kankeleit [3] there is no energy resolution at all, but practically no limit on gamma-ray intensities.

EXPERIMENTAL

We have investigated the 84 keV resonance in ^{170}Yb, the
158 kev resonance in ^{199}Hg and the 60 keV resonance in ^{237}Np
with various methods of gamma-ray detection. These three
transitions differ significantly in the properties of their
gamma-ray spectrum. The following equipment was used:
    (i) "Standard": A 1 3/4" diameter, 1/2" thick NaI(Tl)
scintillator was mounted on a 14 stage photomultiplier
(RCA 8850). The tube base (MWE TB 12)[7] had a transistorized
voltage divider chain for the last dynodes. The high voltage
was set to -1500 V. The proportional pulse was taken from
the 10th dynode and fed via a charge-sensitive preamplifier
to a Tennelec TC 205 A amplifier. Bipolar pulse shaping with
a time constant of 0.25 µs was used. The energy window was
set with an Ortec 406 single channel analyzer.
    (ii) "Fast NaI": The same scintillator, tube and base
as in (i) were used. The high voltage was increased to -2100 V
and pulses were taken from the anode working into a 50Ω resis-
tor. The output pulses were amplified by a factor of 4 without
shaping (e.g. Ortec 574) and then fed into a specially designed
single channel analyzer operating in the 100 mV range (see
Fig. 1).
    (iii) "Plastic": The arrangement is shown in Fig. 2. A
2" thick by 1 3/4" diameter scintillator of 5% tin loaded plastic (NE 140) was mounted on a 14 stage photomultiplier (RCA 8575).
The pulses were again taken from the fast output (anode into 50Ω)
of the MWE TB 12 base at a high voltage setting of -2500V. They
were fed directly into a fast low level discriminator (Canberra
1433) whose output had been modified (-3 V, 0.05 µs) in order to
be compatible with the multiscaler system. The internal deadtime
of the discriminator was set to 70 ns.
    (iv) "Current": A 2" x 1/2" "Integral Line" NaI(Tl) scin-
tillation detector was used. The anode current was converted
into a voltage by the circuit shown in Fig. 3. The integration
time constant of this circuit was set to 20 µs, and a Hewlett-
Packard 2212 A voltage-to-frequency converter was used to
digitize the voltage signal. As discussed in Ref. 3, the additional variance produced by the digitization process can be

reduced by applying an offset voltage to the input of the voltage-to-frequency converter. Since the mean current at the anode is not very stable (e.g. because of temperature variations) the subtraction of an externally generated constant voltage may cause problems. We used an offset voltage proportional to the mean value of the anode current as obtained by an integration with a time constant of about 2s. The output pulses of the voltage-to-frequency converter were fed directly into the multiscaler at a rate of about 10^5 s^{-1}. A typical spectrum is seen in Fig. 4.

All data were stored into 512 memory locations of a PDP 8e computer in double precision. The multiscaler interface contained a buffer register with a capacity of 2^8 counts in order to avoid deadtime losses. The system can be operated at count rates of 3×10^7 s^{-1}. The digital Doppler spectrometer [2] with sinusoidal velocity sweep was run at 40 - 70 Hz, in order to keep the geometry effect small.

RESULTS AND DISCUSSION

Our measurements with the various systems and the three Mössbauer resonances are summarized in Table 1. For the current integration technique, some uncertainties exist with respect to the value of the photon intensity, because it had to be estimated from the statistical scatter of the data, which can be influenced by effects other than counting statistics. Therefore it is difficult to give the exact relative depth of the resonance absorption.

In case of ^{170}Tm the source spectrum contains in addition to the 84 keV resonant gamma-rays, a strong x-ray background around 55 keV and a broad background, probably arising from bremsstrahlung generated by the high-energy electrons from the β decay. The x-ray background was nearly completely removed by a 2 g/cm^2 Cu filter placed at the exit window of the cryostat. Under these conditions the ratio of 84 keV gamma-rays to background radiation is roughly 0.4 and some energy resolution in the detector is of advantage. The fast NaI system therefore gives the best results.

The spectrum of the ^{241}Am source consists practically only of the resonant 60 keV gamma-rays of ^{237}Np. In the Mössbauer setup a small amount of background radiation arises from the Np x-rays emitted from the ^{237}Np absorber. In our setup the absorber contributed only 4% to the integral count rate. As expected, no noticeable difference exists between the fast counting and the current integration method, which both give roughly the same signal-to-background ratio as any energy-selective detection system. Energy resolution is of no advantage in this case and will only slow down the data acquisition. Typical spectra of the ^{237}Np resonance obtained with the plastic scintillation detector can be found in Ref. 8.

The gamma-ray spectrum of ^{199}Au contains a background from the 208-keV radiation, which contributes roughly 10% to the

integral count rate. In this case, fast counting with the plastic
scintillator turns out to be preferable to current integration
for sources of about 1 C or less. One disadvantage is, however,
that the detection efficiency for a 2" thick 5% tin loaded plastic scintillator is only $\sim$ 50% for the 158 keV radiation of ^{199}Hg.
One could probably improve on this by using a 10% lead loaded
scintillator.

SUMMARY

We have shown that with comparatively simple means one can
design a fast pulse handling system based on an uncooled NaI(Tl)
detector. At integral count rates of about 10^6 s^{-1} satisfactory
energy resolution is maintained. At higher count rates the
NaI(Tl) rapidly looses resolution and the single channel analyzer
shows saturation effects. It is then advisable to use a tin or
lead loaded plastic scintillator. This provides only low level
discrimination against noise and x-ray background. However,
since the absorption coefficient for gamma-rays with energies
above $\sim$ 200 keV is very small for loaded plastic, some discrimination against high energy radiation is achieved. The speed
of the system described here is effectively limited by the deadtime of the low level discriminator, which is around 20 ns. It
will operate at count rates of some 10^7 s^{-1}. Current integration
techniques must be employed, when an extremely high flux of
resonant photons is available. The system described here provides an automatic offset and thus reduces the variance arising
from the digitization of the current. The main drawback of the
current integration technique is its loss of direct information
on statistical accuracy and on the signal-to-noise ratio. It
is therefore difficult to obtain the absolute magnitude of the
resonance-absorption and the errors of hyper-fine parameters
may be seriously misjudged in a simple error analysis. Moreover, such systems are not advisable for high-precision measurements of spectral shapes.[3] They should mainly be used for the
measurements of small resonance effects if the photon intensity
at the detector can be made > 10^8 s^{-1}.

One should keep in mind that small statistical errors
reveal other sources of inaccuracies. Most notable are the
uncertainties connected with the curvature of the "off resonance" baseline. These can be reduced by perfect geometrical
alignment and small amplitudes of the drive motion (i.e. a high
drive frequency).

Finally, it should be mentioned, that unfortunately the
most commonly used Mössbauer resonances (^{57}Fe, ^{119}Sn, ^{151}Eu)
are poor candidates for the application of fast counting
techniques. Especially in ^{119}Sn and ^{151}Eu the specific
activities of the source nuclei are too low to make strong
sources. ^{57}Fe is a marginal case. It can be handled better
by a thin tin-loaded plastic scintillator than by current
integration because of the significant high energy background.

Taking the data of [6] as a guide, the application of the plastic scintillator may become worthwhile with a source strength of several 100 mC.

The fast data acquisition systems now available allow the measurement of a resonance absorption strength of the order of 10^{-5} or the determination of spectral shapes with extreme precision. This opens many new possibilities for the application of Mössbauer resonances.

ACKNOWLEDGEMENTS

This work was supported in part by the Kernforschungsanlage Karlsruhe and the Bundesministerium für Forschung und Technologie of the Federal Republic of Germany. The ^{241}Am source was prepared by Dr. J.C. Spirlet of the European Institute for Transuranium Elements, Karlsruhe. The ^{237}NpO$_2$ was given to us by Dr. O. Beck of the Institute of Radiochemistry of the Technical University of Munich.

REFERENCES

1. R.L. Cohen and G.K. Wertheim in Methods of Experimental Physics, ed. R.V. Coleman, Vol. 11 (Academic Press, New York 1974) p. 307
2. N. Halder and G.M. Kalvius, Nucl. Instr. Methods, 108 (1973) 161
3. E. Kankeleit, Proceedings International Conference on Mössbauer Spectroscopy, Cracow 1975, Vol. 2, p. 43
4. W. Müller, H. Winkler and E. Gerdau, J. Physique, C6 (1974) C6-375
5. L.A. Eriksson, C.M. Tsai, Z.H. Cho and C.R. Hurlbut, Nucl. Instr. Methods, 122 (1974) 373
6. J. Becker, L. Eriksson, L.C. Moberg and Z.H. Cho, Nucl. Instr. Methods, 123 (1975) 199
7. Wissenschaftliche Elektronik München
8. W. Potzel, F.E. Wagner, G.M. Kalvius, L. Asch, J.C. Spirlet, W. Müller, contribution to this conference

Table 1: Comparison of results obtained with different detection devices

Source	Absorber	Temp. K Source/Absorber	Detection System	Countrates Integral	Window kHz	Area* %·mm/s	Width* mm/s
^{170}Tm	^{170}YbAl$_3$ in Al	4.2/4.2	standard	120	40	23.1	3.7
		4.2/4.2	fast NaI	800	490	22.3	3.9
		77/77	fast NaI	850	560	6.1	3.7
		77/77	current(NaI)	~1500	–	1.9	3.0
		77/77	plastic	1400	–	2.1	3.6
^{241}Am	^{237}NpO$_2$ metal	77/77	standard	80	60	1.4	2.20
		77/77	fast NaI	550	500	1.1	2.20
		77/77	plastic	1200	–	1.4	2.15
		77/300	plastic	1200	–	0.28(6)	1.80(4)
		77/300	current(NaI)	~1600	–	0.22(5)	1.73(4)
^{199}Au	^{199}HgF$_2$	4.2/4.2	current(NaI)	~6000	–	0.0027(5)	0.75(10)
		4.2/4.2	plastic	1000$^+$	–	0.0033(10)	1.2(3)

*Errors are omitted when smaller than 1%

+Countrates up to ~10^7 s^{-1} would be possible with stronger sources.

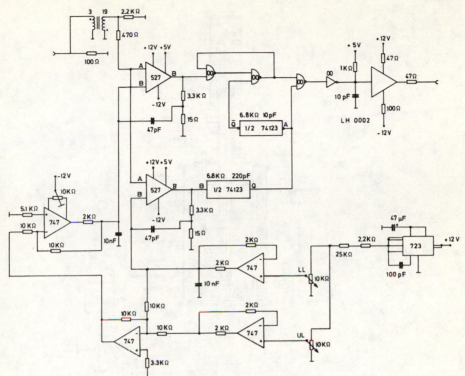

Fig. 1. Circuit diagram of the fast 100 mV input single channel analyzer

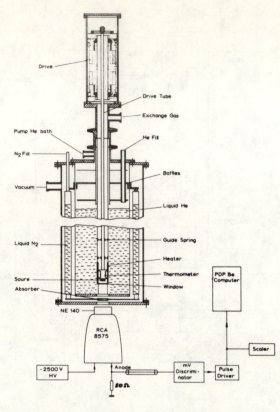

Fig. 2. Experimental arrangement for measurements with the ^{237}Np resonance using a 5% tin loaded plastic scintillator

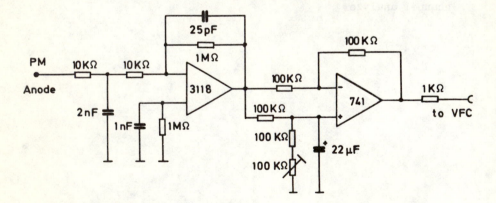

Fig.3. Circuit for the conversion of the anode current of a photomultiplier into a voltage.

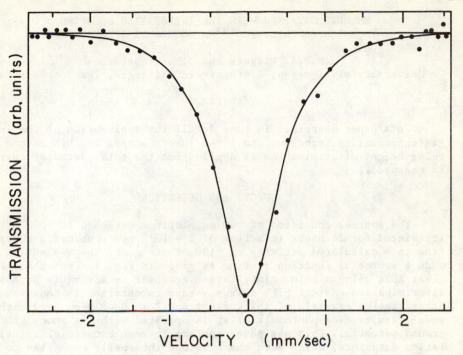

Fig. 4   Mössbauer absorption spectrum of the 158 keV
gamma rays of ^{199}Hg measured by the current
integration method with a source of ^{199}Au in
Pt and an absorber of 10 at% ^{199}Hg in Pd at
4.2 K.

RESULTS OBTAINED WITH THE INTEGRATING COUNTING TECHNIQUE APPLIED TO ^{197}Au MÖSSBAUER SPECTROSCOPY

M.P.A. Viegers and J.M. Trooster
University of Nijmegen, Toernooiveld, Nijmegen, The Netherlands

ABSTRACT

This paper describes in some detail the application of the integrating counting technique[1] to ^{197}Au spectroscopy. In this method no pulse height discrimination is applied but the total detector current is measured.

SOURCES AND DETECTION

The sources consisted of ~40 mg platinum enriched to ~50% ^{196}Pt irradiated for 24 hours in a flux of 2×10^{14} neutrons/cm^2/sec resulting in a calculated activity of ~180 mC of ^{197}Pt. The γ spectrum of such a source as function of time is given in Fig. 1. From this it is clear that using a thin scintillation detector is advisable as some discrimination against high energy γ-rays is obtained in that way. Scintillation crystals of thickness 1 and 2 mm have been used with equal results. The photomultiplier is operated with the anode at ground potential. This may pose problems in some commercial scintillation detectors, which have the envelope internally connected to the kathode, but facilitates the connection to the amplifier low gain of the photomultiplier suffices.

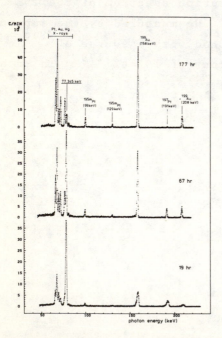

Fig. 1 High resolution γ spectra of enriched ^{196}Pt source at different points of time from the moment, that the sample left the reactor. The spectra were measured with an intrinsic germanium solid state detector with an active area of 200 mm^2 and a thickness of 10 mm (Princeton Gamma Tech IG210U).

As most Mössbauer spectrometers are geared to pulse counting, the output of the photomultiplier is digitized by a voltage to frequency converter and the output pulses of these are then counted in the usual way. This step can be omitted if a signal averager is used. The circuitry for the amplifier/VFC is given in Fig. 2. The photomultiplier gain and the dc level at the input of the VFC are adjusted in such a way that the ac amplitude of the signal does not exceed the voltage range of the VFC (0-10 V). The output pulses of the VFC are counted in the usual way.

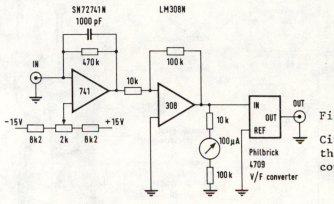

Fig. 2

Circuit diagram of the integrating counting technique

ABSORBER HANDLING

As the integrating counting method results in short measuring times several (~20) absorbers can be measured during the life time of the source. A method was developed to enable quick absorber changes without breaking a vacuum or warming up of the cryostat. The cryostat with insert is shown in Fig. 3a. Source and absorber are contained in an insert, consisting of an 1" diameter thin walled stainless steel tube, which can be evacuated or filled with an exchange gas. The velocity transducer is mounted in the upper part of the insert at room temperature. On top of the insert, outside the vacuum region, a Michelson-type interferometer is placed for absolute velocity measurements. To drive the source a transducer extension rod is used, made from a $\frac{1}{2}$" diameter thin walled stainless steel tube. At the lower end of the insert the extension rod is held in place by a phosphor-bronze leaf spring, electroplated with 50 μ of copper, which serves also as a thermal connection with the helium bath.

This system has a great versatility, because at the end of the insert various tail-sections can be mounted. Fig. 3b shows the tail-section designed for measurements at temperatures between 1.2°K and 100°K.

Vacuum grease (Apiezon N) is used to glue the source to an aluminium source holder, which is subsequently mounted at the end of the transducer extension rod. The upper part of the insert is then lowered on the tail-section, which has been placed in the lead shielded lock on top of the cryostat. Guide pens align both parts, so that

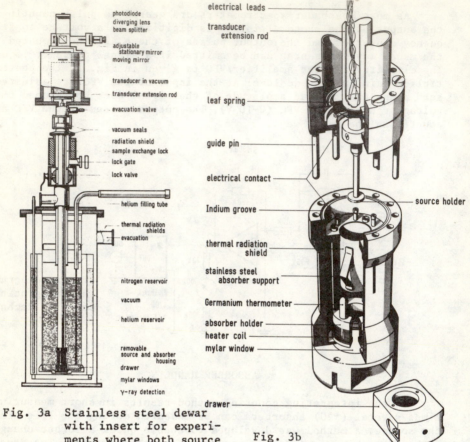

photodiode
diverging lens
beam splitter

adjustable
stationary mirror
moving mirror

transducer in vacuum

transducer extension rod

evacuation valve

vacuum seals
radiation shield
sample exchange lock
lock gate
lock valve

helium filling tube

thermal radiation
shields
evacuation

nitrogen reservoir

vacuum

helium reservoir

removable
source and absorber
housing
drawer
mylar windows
γ-ray detection

electrical leads

transducer
extension rod

leaf spring

guide pin

electrical contact

Indium groove

source holder

thermal radiation
shield

stainless steel
absorber support

Germanium thermometer

absorber holder
heater coil
mylar window

drawer

Fig. 3a Stainless steel dewar
with insert for experiments where both source
and absorber are cooled.

Fig. 3b
Tail-section of the insert.

electrical connections are made by six gold-plated contacts. The two
parts are bolted together with stainless steel screws and an indium
ring provides the vacuum seal.

For measurements at 4.2°K or lower the absorber is mounted in a
holder outside the vacuum region in direct contact with the coolant
liquid. To change the sample the insert is pulled out of the helium
bath until the tail-section reaches the sample exchange lock, where
the valve is closed and the holder pulled out. Then another holder,
precooled in liquid nitrogen is pushed in, the valve is reopened and
the insert is slowly lowered in its original position. The whole
operation can be executed in less than 5 minutes and consumes less
than 100 cm^3 liquid helium.

For temperature dependent measurements above 4.2°K the absorber
is wrapped in aluminium foil to minimize temperature differences over
the sample. The absorber is then mounted inside the evacuated region
where it can be heated with a heater coil. A copper tube with aluminium bottom, anchored at 4.2°K, provides a thermal shielding of the

source. The temperature of the absorber is measured with a germanium sensor and regulated with a BOC model 3010 temperature controller. Stabilization better than 0.05°K could be obtained.

INTENSITY CALIBRATION

The signal measured as function of the velocity v is given by:

$$I(v) = I_o [1 - \Delta(v)] [1 - L(v)] + I_d \quad . \tag{1}$$

I_o is proportional to the γ-ray intensity detected when the source is in equilibrium position, I_d is given by the dc level potentiometer of the preamplifier, $L(v)$ is the Mössbauer absorption and $\Delta(v)$ is the intensity variation due to the variation of the source-collimator distance. For a constant acceleration spectrometer v is proportional to time t during each half period $\Delta(v)$ is given by:

$$\Delta(v) = \Delta(t) \propto \frac{2x(t)}{x_o} \quad , \tag{2}$$

where x_o is the equilibrium position of the source and $x(t)$ the displacement of the source from equilibrium position. In general $\Delta(t)$ ≪ 1 and similarly $L(t) \ll 1$, hence

$$I(v) \simeq I_o + I_d + I_o (1 - \Delta(t) - L(t)) \quad . \tag{3}$$

To determine the true relative absorption intensity $L(t)$, I_o should be known whereas $I_o + I_d$ is measured. However I_o can be determined from the measured base line curvature $I_o \cdot \Delta(t)$ when $\Delta(t)$ and x_o are known. For constant acceleration with period T:

$$x(t) = + v_m [t - \frac{2}{T} t^2] \quad , \tag{4}$$

where the sign is different for the two half periods and v_m is the maximum velocity.

MEASUREMENTS IN A MAGNETIC FIELD

For a few measurements an external magnetic field was applied with a superconducting solenoid placed at the bottom of the helium vessel (not shown in Fig. 3).

The photomultiplier tube was placed outside the influence of the magnet using a bend light-pipe made of a 5 cm diameter lucite rod with a length of 2 m. At one end the 2 mm thick scintillation crystal was mounted and at the other end the photomultiplier tube. The gain and resolution of this detector were smaller by a factor of about 4 compared to measurements without light-pipe. However, this did not significantly influence the measurements, when the integrating counting technique was used.

DISADVANTAGES

Recently Kankeleit[2] has given a thorough discussion of the application of current integration methods. According to this author there are two drawbacks of this method. First of all scintillation detectors may show several decay times leading to distortion of the absorption signal. In our experiments the intensities and intensity variations are considerably lower than in Kankeleits experiments and no adverse effect could be detected in the Mössbauer spectra. There is however a small phase shift of the signal of the order of 4°. This results in an incorrect IS. However as the shift is opposite in sign for the two half periods of motion, averaging results in the correct

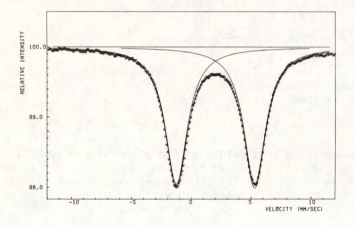

Fig. 4   Spectrum of gold complex containing 133 mg/cm^2 Au.
         Measuring time 90 minutes, started 20 hours after end of
         source irradiation.

IS. The isomer shift of a gold foil with respect to the Pt-source is found to be -1.194 ± 0.005 mm/sec independent of the velocity range. A second disadvantage is the low pass dc coupling of the detector. Even very weak coupling of for instance the drive signal to the transducer with the signal amplifier will result in a signal in the spectrum. Spurious signals, which could be correlated with oscillations on the velocity signal, were sometimes observed. If however the velocity is a smooth function of time no effect is seen in the spectrum, as can be seen in the spectrum of Fig. 4.

REFERENCES

1. M.P.A. Viegers and J.M. Trooster, Nucl. Instr. Meth. 118, 257 (1974).
2. E. Kankeleit, Proceedings International Conference on Mössbauer Spectroscopy, Vol. 2, Cracow (1975).

SPHERICAL ELECTROSTATIC ELECTRON SPECTROMETER FOR MÖSSBAUER SPECTROSCOPY*

N. Benczer-Koller and B. Kolk[†]
Department of Physics
Rutgers University, New Brunswick, N. J. 08903

ABSTRACT

A high transmission spherical electrostatic electron spectrometer has been constructed for combined Mössbauer and conversion electron spectroscopies. To date, a transmission of 7% and an energy resolution of 2.5% at 14 keV have been achieved for a source of 1 cm diameter.

INTRODUCTION

It may prove advantageous in Mössbauer spectroscopy to detect the internal conversion electrons emitted in the de-excitation of the resonantly excited state. The internal conversion coefficients for most Mössbauer transitions are very large, so that counting rates may be increased many fold over the gamma ray rates. Resonant detectors[1] were proposed many years ago and have been used very successfully since; conventional electron counters[2] or spectrometers[3,4,5] have also been used for particular applications.

Certain investigations, however, require a direct measurement of the configuration of the atomic electrons and consequently demand the use of spectrometers with high transmission as well as high resolution. For example, isomer shifts measure the total charge density of s electrons at the nucleus,

$$\text{Isomer shift} \sim \sum_n |\psi_{ns}(o)|^2 \, .$$

By contrast, by internal conversion techniques it is possible to select conversion electrons from each ns shell and measure separately the charge densities $|\psi_{ns}(o)|^2$ of the core and conduction electrons at a chosen nucleus embedded in a given host and thus achieve a better understanding of the ion-host interaction.

A second area of interest is related to a microscopic understanding of the hyperfine fields acting at nuclei in magnetic materials. Again in this case, the total hyperfine field may be measured by a variety of

*Supported in part by the N.S.F.
[†]Present address: Boston University, Boston, Mass.

techniques but only the sum of many contributions is observed:

$$\text{Hyperfine field} \sim \sum_n \{ |\psi^{\uparrow}_{ns}(o)|^2 - |\psi^{\downarrow}_{ns}(o)|^2 \} .$$

By means of Mössbauer-conversion electron spectroscopy the spin polarization of each s electron shell may be measured separately,[6] in order to elucidate the relative contributions to the hyperfine field arising from core or conduction electron polarization.

EXPERIMENTAL REQUIREMENTS

In these experiments the state of interest is excited by Mössbauer spectroscopy and the decay electrons are detected with an electron spectrometer capable of resolving the energies of the various s-shell electrons from each other.

For example (Table I), in the case of ^{57}Fe an energy resolution of better than 5% is necessary to resolve the core electrons from each other, but a resolution better than 0.5% is required to select the conduction electrons.

Table I. Energies of the conversion electron emitted in the decay of the 14.4 keV state of ^{57}Fe.

Electron Configuration	Electron energy	(keV)	ΔE(keV)	ΔE/E
1s(K)		7.27		
2s(L$_I$)		13.54	0.748	5.25%
3s(M$_I$)		14.29		
4s(N$_I$)		14.38	0.093	0.85%

Besides energy resolution, and high transmission, the requirement of combined Mössbauer conversion electron spectroscopies demands that the source of electrons which acts as the absorber to be resonantly excited by the recoilless radiation be large, a difficult condition to satisfy in most electron spectrometers.

ELECTROSTATIC ELECTRON SPECTROMETER

A high resolution, high transmission, spherical electrostatic condenser was built to satisfy the requirements listed above. This spectrometer (Fig. 1) is a scaled down version of a much larger instrument designed and built by Ritchie and Birkhoff[7] at Oak Ridge.

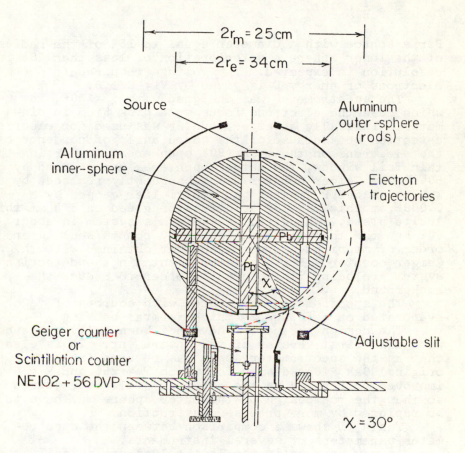

Fig. 1. Schematic diagram of the electron spectrometer.

An inner sphere mostly filled with Pb is surrounded by a spherical cage constructed from rods in order to reduce electron scattering. The source is mounted on the inner sphere. The electron trajectories are defined by an adjustable exit ring slit subtending an angle χ to the center of the inner sphere. The resolution R and transmission T of the spectrometer are entirely determined[7] by this angle, and for a point source are given by

$$R = \frac{\Delta E}{E} = \frac{1}{16} \chi^2$$

$$T = \frac{1}{4} \chi$$

For a source with a diameter equal to 15% of the radius of the inner sphere a deterioration of less than 20% in resolution is expected. The voltage required to focus electrons of energy E is given by $V = E/1.6$.

The spectrometer was designed with $\chi = 30°$, in which case the electrons focus on a disk 5 cm in diameter. Initially a Geiger counter was used for electron detection but because of the sharp angle of incidence of the electrons on the mesh (90% open area) supporting the thin foil window, the actual open area was reduced by 50%. The Geiger counter was subsequently replaced by a low noise 56DVP Amperex photomultiplier capped by a 0.0005 cm thick NE 102 scintillator glued on a 5 mm thick lucite plate. The efficiency of this system is about 90% but the thermal noise (about 20 counts/sec) is an order of magnitude larger than that obtained with the Geiger counter. Further improvements in the detection system are being implemented in order to reduce the background.

The spectrometer was tested with sources of ^{57}Co evaporated on a 50 $\mu g/cm^2$ thick formvar backing.

The design parameters have not been fully achieved with this first version of the instrument. It is clear that as the spectrometer scale was reduced from the original Oak Ridge design for which T = 25% and R = 6%, improved uniformity in the electric field is required so that the rod design of the outer sphere may have to be replaced by more precise construction.

Table II shows a comparison between the spectrometer parameters of several instruments.

As may be seen in the last column of Table II, the luminosity of the spherical condenser is considerably larger than that of the magnetic solenoid used in earlier experiments. This extremely large luminosity coupled with the relatively simple design is the main advantage of this spectrometer and makes it a unique tool for precision measurements of charge or spin densities of s electrons at select nuclei embedded in a variety of hosts.

Table II. Comparison of the performance parameters of various spectrometers. The luminosity is the product of the source area and the transmission.

Spectrometer	Source Diameter	$\frac{\Delta E}{E}$(%)	T(%)	Luminosity
Long lens magnetic solenoid: Rutgers	2 mm	4	2	0.03
Spherical electrostatic condenser:				
Oak Ridge		6.3	25	
Rutgers (design)	<1 cm	1.5	13	1.3
Rutgers (observed)	1 cm	2.5	7	0.75

REFERENCES

1. K. P. Mitrofanov, N. V. Illarionova, V. S. Shpinel Pribory: Tekhn. Eksperim 8, 49 (1963) [Engl. Transl. Instr. Exptl. Tech. (USSR) 3, 415 (1963)].
2. K. R. Swanson and J. J. Spijkerman, J. Appl. Phys. 41, 3155 (1970); Y. I. Sozumi, D. I. Lee, and I. Kadar, Nucl. Instr. Meth. 120, 23 (1974).
3. E. Moll, E. Kankeleit, Nucleonik 7, 180 (1965); H. Bokemeyer, K. Wohlfahrt, E. Kankeleit, and D. Eckardt, Z. Physik A274, 305 (1975).
4. V. Bävestram, C. Bohm, B. Ringström, and T. Ekdahl, Nucl. Instr. Meth. 108, 439 (1973); V. Bavestram, T. Ekdahl, C. Bohm, B. Ringstrom, and D. Liljequist, Nucl. Instr. Meth. 115, 373 (1973).
5. M. Fujioka and T. Shinohara, NIM 120, 547 (1974); T. Shinohara, M. Fujioka, H. Onodera, K. Hisatake, H. Yamamoto, and H. Watanabe, private communication.
6. C. J. Song, J. Trooster, N. Benczer-Koller, and G. M. Rothberg, Phys. Rev. Lett. 29, 1165 (1972)- C. J. Song, J. Trooster, and N. Benczer-Koller, Phys. Rev. B9, 3854 (1974).
7. R. H. Ritchie, J. S. Cheka, and R. D. Birkhoff, Nucl. Instr. Meth. 6, 157 (1960); R. D. Birkhoff, J. M. Kohn, H. B. Eldridge, and R. H. Ritchie, Nucl. Instr. Meth. 8, 313 (1960).

RAYLEIGH SCATTERING OF MOSSBAUER RADIATION FROM PLASTIC CRYSTALS

D. C. Champeney
University of East Anglia, Norwich, England

ABSTRACT

A survey is given of some results on the Rayleigh scattering of Mossbauer radiation from plastic crystals in polycrystalline form. In each case the elastic component in the scattered radiation shows a more or less abrupt decrease in value on warming through each phase transition at which molecular motion increases. The possibility of obtaining detailed information on the nature of molecular rotation in the plastic crystal phase is discussed.

SURVEY

Experiments by D. F. Sedgwick (unpublished) have been performed on a number of plastic crystals in polycrystalline form. In some of these the recoilless scattering fraction fell abruptly to zero on warming through the brittle to plastic phase transition: in this category were cyclopentane, pivalic acid, and 2 Methyl 2 propanethiol, as well as the already published case of cyclohexane. Heat capacity data show that the entropy change at the brittle to plastic transition in these cases is large compared with the entropy change on melting. In other cases the recoilless fraction only showed a discontinuity, so that a measurable recoilless fraction existed in the plastic crystal phase. Thianaphthene was one such case, there being a small drop at the brittle to plastic phase transition, and 2 chloro 2 methyl propane was another similar case. In this latter compound there are two phase transitions below the melting point, and the recoilless fraction fell to a finite value at the lower of these and then to zero at the higher one. In each case the magnitude of the drop in the recoilless fraction shows a close correlation with the corresponding entropy change, derived from heat capacity data, though due to the complexity of the molecules further exact calculation appears difficult.

Experiments carried out by E. S. M. Higgy (to be published) on nondecane and adamantane also show discontinuities in recoilless fraction at the brittle to plastic phase transition, but in these cases, due to the simplicity of the molecules, there does seem to exist the possibility of some numerical evaluation of the results in terms of different models of molecular motion. The backbone of nondecane consists of a zig-zag chain of carbon atoms, and if this simply rotated about its long axis in the plastic phase, each individual carbon would rotate in a circle of radius 4.2×10^{-11} m.   Adamantane is a 'globular' molecule, $C_{10}H_{16}$, with its carbon atoms arranged so that if it performs hindered rotation it will appear as though only four of the carbons move at each step, due to the symmetry of the molecule, and again the effect of such a motion should be calculable in detail from the known dimensions of the molecule.

REFERENCE

1.   D. C. Champeney and D. F. Sedgwick, J. Phys. C, 4,2220 (1971).

MÖSSBAUER SCATTERING FROM A SUPERCOOLED LIQUID

M. Soltwisch, M. Elwenspoek,and D. Quitmann

Institut für Atom- und Festkörperphysik, Freie Universität Berlin, Boltzmannstraße 20, D-1000 Berlin 33, Fed.Rep. Germany

ABSTRACT

By a coherent Mössbauer-scattering experiment, the dynamical structure factor $S(k,\omega)$ was determined for pure liquid glycerol. As a function of temperature ($-30^{\circ}C \leq T \leq 0^{\circ}C$) and momentum transfer ($0.7\text{Å}^{-1} \leq k \leq 4\text{Å}^{-1}$) the width and intensity of the quasielastic line and the intensity of the inelastic line were measured. From the ratios of the intensities the mean-square vibrational amplitude of the molecules was derived while from the linewidth of the quasielastic part we extracted translational and rotational diffusion parameters.

INTRODUCTION

An experiment to study quasielastic non-resonant scattering of photons was performed, which employs the very high energy resolution of Mössbauer-γ-radiation. The first system studied was pure supercooled glycerol, a liquid showing slow translational and rotational diffusion of the molecules. The scattered intensity was registered as a function of temperature ($-30^{\circ}C \leq T \leq 0^{\circ}C$) and momentum transfer ($0.7\text{Å}^{-1} \leq k \leq 4\text{Å}^{-1}$). Three experimental parameters were obtained at every (k,T)-point:

a) the diffusional broadening of the sharp Mössbauer line (fitted by assuming a Lorentzian energy shape of the quasielastic scattering);

b) the ratio of quasielastic to inelastic intensity;

c) the total intensity.

Except for the energy distribution of the inelastic part one measures the dynamical structure factor $S(k,\omega)$ of the liquid system, a quantity otherwise obtainable only by coherent neutron scattering.

EXPERIMENTAL DETAILS AND LIMITATIONS OF THE METHOD

A 250 mCi ^{57}Co(Rh)-source[1] (1 mm x 6 mm x 6 μ), inside a brass and lead collimater, a double walled, temperature regulated scattering chamber in Laue-arrangement (transmission scattering) with a sample thickness of 5 mm, and a collimated detector system with Mössbauer absorber and Si(Li)-diode were the main parts of the experimental setup

They were mounted on a conventional horizontal X-ray gonio-
meter. Around its vertical axis, source and sample could
be rotated independently. The scattered intensities were
corrected for white background and Compton-scattering using
pulse height spectra taken at different angles.

Without the Mössbauer absorber the apparatus constitutes
an X-ray scattering arrangement, where the total intensity
scattered at angle Θ is proportional to the static struc-
ture factor $S(k)$ with $k=4\pi\sin\Theta/\lambda$ and $\lambda=0.86\text{Å}$. The wavelength used here allows the observation of the next neigh-
bour correlations in condensed systems at moderate angles;
one obtains a broad structure maximum from a liquid or
sharp Bragg peaks from a single crystal. The occurrence of
well defined maxima at $\Theta\neq0$ is an effect of the coherence
of the scattering process. This is different from normal
Mössbauer-absorption or -emission experiments as they are
normally done in studying properties of condensed systems.
There one has individual absorption or emission which are
perfectly incoherent processes and one has only one fixed
value of k, $k=2\pi/\lambda$. Mössbauer scattering now allows one to
study correlation between scatterers. For $1/k\approx$molecular
distance the response of the system stems to a great part
from the correlated arrangements and motions of the scattering sites, while at large momentum transfers k one again
observes only the individual molecules or atoms.

We consider cases where the motions of the molecules can
be divided into an oscillatory (spatially bounded) part of
energy $k_BT\approx20\text{meV}$ (phonons) and a spatially unbounded portion (recoilfree but Doppler-shifted scattering due to the
velocity distribution of the diffusing molecules) producing an energy change of $\leq10^{-7}\text{eV}$. Mössbauer-scattering
distinguishes the scattering events due to both parts and
the latter, quasielastic part can be energy-analysed. The
ratio between the two contributions is governed by a
Debye-Waller-factor $f=\exp(-k^2\langle x^2\rangle)$, where $\langle x^2\rangle$ is the mean
square vibrational amplitude of the individual molecules;
here the coherence of the scattering has to be taken into
account.

The energy range of $\leq10^{-7}\text{eV}$ for the quasielastic part
restricts the applicability of Mössbauer-scattering in
liquids to the highly viscous regime. Glycerol is a standard subject in this field and has often been studied by
other methods like NMR, NQR, dielectric relaxation, Raman-
scattering etc.

RESULTS

Fig. 1 shows the line-broadenings W_m as a function of the
scattering angle at three temperatures. Fig. 2 gives the

measured fraction f_m of quasielastic intensity to total intensity for the same temperatures. Fig. 3 shows the inelastic and the total intensity, I_{in} and I_{tot}, respectively for one temperature (-16.5 °C). The structure maximum corresponds to the mean molecular distance of ~5Å. The line drawn for I_{tot} is from a Percus-Yevick hard sphere calculation with an additional fast and independent vibration of the molecules. From the last assumption follows

$$I_{tot} = I_q + I_{in} = f \cdot S(k) + (1-f) \cdot S_s(k)$$

where $S_s(k)$ is the self-part of $S(k)$ and $S(k)$ is the static structure factor for $<x^2>=0$.

In fig. 2, the experimental values $f_m=I_q/I_{tot}$ and the calculated values $fS(k)/(fS_d(k)+S_s(k))$ are plotted for the three temperatures (note that $S(k)=S_s(k)+S_d(k)$); the agreement is fairly good.

The $<x^2>$-values derived in this way show a strong increase with temperature (larger than in the harmonic approximation for the solid where $<x^2>\alpha T$). This behavior was already earlier measured by Champeney[2].

A model which can describe the quasielastic linewidths given in fig. 1, must include translational as well as rotational diffusion. The latter contributes to the linewidth due to the very unspherical shape of the glycerol molecule. The coherence of the scattering process is essential because it creates a very different k-dependence for the translational and rotational contribution to the linewidths[3]. We assume jump models for both diffusion processes with the same correlation time τ; the mean jump widths are $<r^2>$ for the translation and $<\varepsilon^2>$ for the rotation. These three quantities are determined from a fit to the linewidth as a function

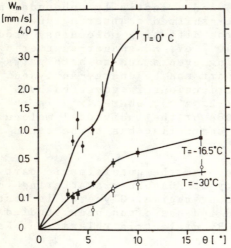

Fig. 1. Broadenings of the quasielastic scattered intensities as a function of the scattering angle Θ, $k=4\pi \sin \Theta /\lambda$. Results:[$\tau$ in 10^{-7}s]
$T =-30.0°C$: $\tau = 1.0$ (2), $<r^2> = 0.10...1.0$ Å^2, $<\varepsilon^2> = 0.1 ...1.0$rad^2
$T =-16.5°C$: $\tau = 0.4$ (2), $<r^2> = 0.80...1.3$ Å^2, $<\varepsilon^2> = 0.05...0.4$rad^2
$T = 0.0°C$: $\tau = 0.07(2)$, $<r^2> = 0.80...1.3$ Å^2, $<\varepsilon^2> = 0.25...0.7$rad^2

k. These results are given in the capture of fig. 1. From these, the diffusion constants $D_{trans} = <r^2>/6\tau$ and $D_{rot} = <\varepsilon^2>/6\tau$ can be derived; their temperature dependence follows an Arrhenius law. The parameters of translational motion can be determined considerably better than the rotational ones. For a more detailed discussion of parts of the results see[4].

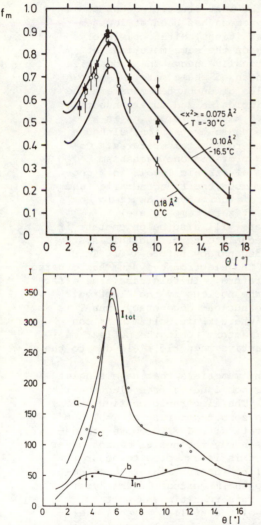

Fig. 2. Ratio of quasielastic to total intensity as a function of scattering angle Θ for three temperatures.

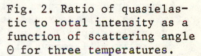

Fig. 3. Total and inelastic scattering intensity $I_{tot} = I_q + I_{in}$ and I_m, respectively. Points: experimental results (T=-16.5°C). Curve a gives I_{tot} and curve b is I_{in}, calculated with $<r^2> = 0.1Å^2$. Curve c: $I_{tot} = I_q$ calculated for $<x^2> = 0$. The intensities are corrected for background and Compton-scattering.

1. Radiochemical Centre, Amersham, England.
2. Champeney, D.C., and Woodhams, F.W.D., J. Phys. B, 1, 620 (1968).
3. Sears, V.S., Can.J.Phys., 44,1299 (1966).
4. Soltwisch, M., Elwenspoek, M.,Quitmann, D., appears in Molecular Physics 1977.

118

Mössbauer Studies in a ^{3}He-^{4}He Dilution Refrigerator:
Search for "Nuclear Cooperative Phenomena"[*]

B. B. Triplett, N. S. Dixon, Y. Mahmud, and S. S. Hanna

Department of Physics, Stanford University, Stanford CA

Historically, one of the most fruitful areas to
search for new physical phenomena has been at low temperatures. Cooperative transitions, often completely
unexpected from experiments on the same materials at
higher temperatures, dramatically change the properties
of some of the simplest condensed phase materials known
to science. Classic examples of such phenomena are the
superfluid transition in liquid ^{4}He and superconducting
transitions in many metals. Magnetic ordering is also
an example of a cooperative phase transition, although
one not exclusively confined to low temperatures. Cooperative magnetic ordering provides one mechanism
whereby a set of degenerate electronic levels in a condensed phase material may remove their degeneracies and
thus satisfy the third law of thermodynamics (zero entropy, or zero degeneracy, at zero temperature).

The nuclear ground state will also be degenerate if
the nuclear spin I is not equal to zero, and this system
generally will remove its degeneracy before T=0. This
happens quite naturally if the electronic system has ordered
magnetically at high temperature. In this case, the effec-
magnetic field applied to the nucleus by the electronic
system begins to remove the nuclear degeneracy when T is
lowered to $T \approx g_N \mu_N H_{eff}/k_B$ where k_B is Boltzmann's constant, g_N is the nuclear g factor, μ_N is the nuclear mag-
neton, and H_{eff} is the effective magnetic field due to the
magnetic electronic system.

However, there are many materials that have singlet
electronic ground states or nonmagnetic free electron-like
electronic configurations. The electronic configurations
in such materials will have zero degeneracy at T=0, but
how will a nuclear groundstate with $I \neq 0$ remove its degeneracy? There is no reason why the nuclei cannot order
magnetically and, in fact, this is expected to occur
through the magnetic hyperfine interaction at some temperature T_c proportional to the magnitude of the hyperfine interaction squared. This result means that T_c is
proportional to the square of the nuclear moment, whereas, the analogous T_c for the degenerate electronic system is proportional to the square of the electronic
moment. Since the magnitudes of nuclear moments are $\approx$
10^{-3} the typical electronic moment effective nuclear-
nuclear magnetic order is expected to occur at temperatures $\approx 10^6$ smaller than typically observed for conventional electronic ordering. Closer inspection[1] of the
full Hamiltonian for the combined electronic-nuclear

system in such singlet ground state materials leads to
a somewhat different characterization of the system.
It is convenient to introduce the parameter
$\eta = 4J(0)\alpha^2/\Delta$ where $4J(0)\alpha^2$ represents the strength of
the magnetic exchange interaction and Δ represents the
splitting between the two lowest electronic singlet
states. We can now identify three regimes of magnetic
ordering:

(a) The region $\eta \gg 1$ corresponds to an exchange interaction much larger than the splitting between the two
electronic singlets, consequently, the levels behave
as if they are "degenerate" and order magnetically in
the usual fashion observed for strictly degenerate
levels.

(b) The region $\eta \ll 1$ corresponds to such a weak exchange interaction that only nuclear-nuclear cooperative
ordering can remove the nuclear degeneracy. The electronic singlets do not order.

(c) The region $\eta \lesssim 1$ corresponds to an interesting
intermediate region. If the nuclear part of the
Hamiltonian is neglected, the electronic system will not
order if $\eta < 1$. However, diagonalization of the full
electronic-nuclear Hamiltonian shows that in the region
$0.7 \lesssim \eta \lesssim 1$, the system orders through a mechanism that
can be described as electronic-nuclear cooperative
ordering with T_c proportional to the product of the
electronic moment and the nuclear moment. Transition
temperatures found by diagonalizing the full Hamiltonian
appropriate to a system containing [141]Pr are shown as
solid curves in Fig. 1. The solutions obtained with
only the nuclear part of the Hamiltonian (for $\eta < 1$) and
the electronic part of the Hamiltonian (for $\eta \geq 1$) are
shown as dashed curves. The important point is that
for $\eta = 0.9$ for instance, transition temperatures are
enhanced by an order of magnitude over those expected
for effective nuclear-nuclear cooperative ordering
(i.e., neglecting the role played by the electronic
system). Such transition temperatures are easily within
the range of current dilution refrigerators and evidence
for this phenomenon has already been reported in PrCu$_2$.[2]
We have developed a program at Stanford to look for such
systems as a part of a larger low temperature Mössbauer
program. Moreover, as shown in Fig. 2, the full
Hamiltonian also predicts that fairly dramatic enhance-
emnts of the electronic moment $\langle J_z \rangle$ in electronically
ordered systems can occur at low temperatures if η is
only slightly larger than 1.

Although we have studied materials containing both
[141]Pr and [169]Tm in our dilution refrigerator, we have
made more rapid progress with the latter Mössbauer transition. Figure 3 shows a schematic diagram of the
^{3}He-^{4}He dilution refrigreator adapted for experiments

with the ^{169}Tm transition. The absorber is attached to a
refrigerator mixing chamber. The source of the 8.4 keV
radiation of ^{169}Tm is ^{169}Er produced by neutron irradiation of a foil containing about 8% by weight ^{168}Er in Al.
A sinusoidal Doppler modulation of the γ-ray energy is
produced with a modified version of a commercially available electromechanical transducer which easily reaches
the required velocities of ± 70 cm/sec (with less than
1% deviation from sinusoidal form). The source, attached
to the drive rod of the transducer, is maintained at a
temperature of about 140 K at a position about 4 cm from
the absorber. The γ radiation passes through seven Be
windows (three of which are vacuum seals) and is detected

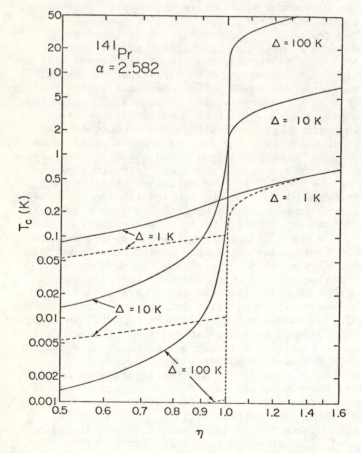

Fig. 1. Transition temperature T_c as a function of $|\eta|$ for
Δ = 1, 10, and 100 K.

by a large window Xe-CO_2 proportional counter mounted on the outside of the dewar. Counting rates, even after the strong attenuation produced by $\approx$ 6-8 mg/cm^2 ^{169}Tm, were of order 90,000 per minute in the γ-ray window. Heating from the γ-ray flux has not been a significant problem in absorber experiments.

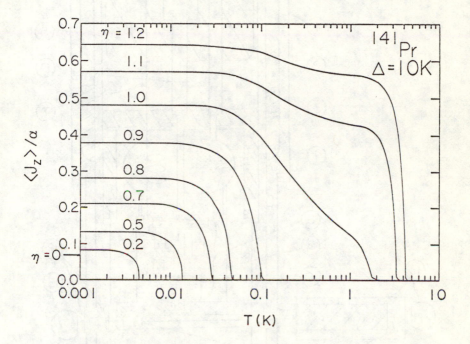

Fig. 2. Temperature dependence of $\langle J_z \rangle /\alpha$ (which is proportional to the effective magnetic hyperfine field) calculated in the molecular field approximation for various values of η with Δ = 10 K and α = 2.582.

The first stage of our program has involved the measurement of the hyperfine interactions in Tm compounds known to order magnetically. Examples of these measurements, at temperatures sufficiently low so that only the electronic ground state is occupied, are shown in Figs. 4-6 and a summary of the corresponding hyperfine interactions is given in Table 1. Figure 4 shows a Tm metal absorber produced by rolling a Tm metal ingot to a thickness of 10 mg/cm^2. This technique produced an absorber with strong alignment of the polycrystallites,

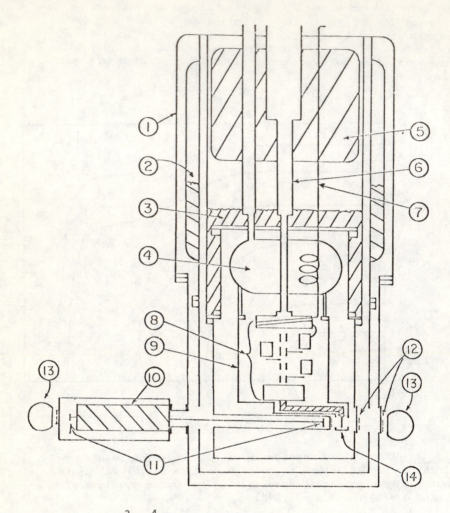

Fig. 3. The ^{3}He-^{4}He dilution refrigerator cross section:
(1) outer shell, (2) liquid nitrogen reservoir with LNT
shield attached below, (3) liquid helium reservoir with
HeT shield attached below, (4) pumped helium reservoir
(1.2 K), (5) plastic foam convection baffle, (6) pumping
line of still, (7) ^{3}He return line, (8) refrigerator
assembly, (9) 1.2 K thermal shield, (10) velocity transducer and Moiré interferometer, (11) ^{169}Er(Al) source and
^{57}Co(Pt) calibration source, (12) Be windows, (13) Xe-CO$_2$
proportional counter, (14) absorber radiation shield (Be
windows not shown).

Table 1a: RECENT ^{169}Tm MAGNETIC HYPERFINE STRUCTURE MEASUREMENTS AT STANFORD[†]

Material	ΔE_{tot} (cm/sec)	Internal field (MG)	f_{mag}[†††]	QI[††] (cm/sec)	f_{elec}	Reference
$TmFe_2$: Tm^{3+}	116.5	7.03	1.	7.84	1.	Ref. 5
Tm : Tm^{3+}	107.7(1)	6.50	0.925	6.99(2)	0.89	This work
$TmAl$: Tm^{3+}	111.0(8)	6.70	0.952	6.71(6)-	0.856	This work
$TmAl_2$: Tm^{3+}	91.0(6)	5.49	0.781	5.55(5)	0.708	This work
TmS : Tm^{3+}	≈ 48.25	2.91	0.414	≈ 1.00	0.128	This work
$TmSe$: $\approx 80\%$ Tm^{3+}	≈ 32.06	1.93	0.275	≈ 0.40	0.051	This work
$TmTe$: Tm^{2+}	57.2(6)	3.45	0.82	6.91(8)	0.82	This work

Table 1b: ESTIMATED $Tm^{3+}(4f^{12})$ AND $Tm^{2+}(4f^{13})$ FREE ION PARAMETERS

	Internal field (MG)	QI[††] (cm/sec)	Reference
Tm^{3+} ($^3H_6$) free ion	6.93	(7.84)	Ref. 6
Tm^{2+} ($^2F_{7/2}$) free ion	4.23	(8.43)	Ref. 6

[†] The errors in Table 1a are statistical only. We have not yet achieved velocity stability much better than ± 0.5%, and we do not claim reproducibility better than this. The materials TmS and TmSe show spectra which must be fitted by a distribution of magnetic and electric quadrupole interactions. We report the average value of the distribution.

[††] $QI = e^2qQc/4E_\gamma$, where E_γ = 8.401 keV.

[†††] The values f_{mag} and f_{elec} are our best estimates for the fraction of the magnetic and electric free ion interactions, respectively.

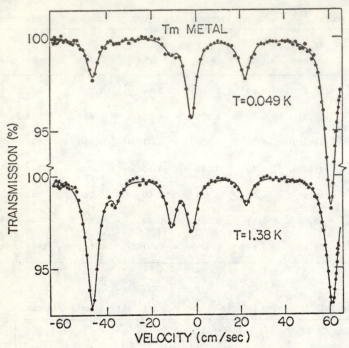

Fig. 4. Mössbauer spectra of partially oriented ^{169}Tm foil at 0.049 K and 1.38 K. The foil was produced by rolling a Tm metal ingot to a thickness of $\approx$ 10 mg/cm^2. This procedure resulted in partial orientation of the polycrystallites in the foil and the observed relative transition intensities are 3:0.65:1:1:0.65:3 instead of the usual 3:2:1:1:2:3 appropriate for random orientation of the polycrystallites.

as is evident from the suppression of the $\Delta m_I = 0$ transitions in the spectra. Cooling this absorber to 0.049 K produced strong nuclear polarization. No increase in the hyperfine field was observed nor is any significant increase expected since the field is near its maximum value and is thus insensitive to the hyperfine enhanced mechanism. However, in Figs. 5 and 6, we show results on some cubic Tm^{3+} compounds (TmAl$_2$, TmS, and TmSe) having singlet ground states and hyperfine fields substantially reduced from the Tm^{3+} free ion values. These materials are expected to show a gradual enhancement of the hyperfine field when the temperature is lowered to $T \approx g_N \mu_N H_{eff}/k_B$, however, to date, we have observed no such enhancement. Some asymmetries are apparent in the TmS and TmSe spectra (particularly noticeable in lines #2 and #6), but these effects are temperature independent and can be explained by a narrow distribution of strain-induced quadrupole interactions. Figure 6 also shows one of our spectra taken on TmTe, noteworthy because it is the first measurement of Tm^{2+} hyperfine structure (a small residual contaminant which did not order magnetically is indicated by the dashed curve above the TmTe spectrum in Fig. 6).

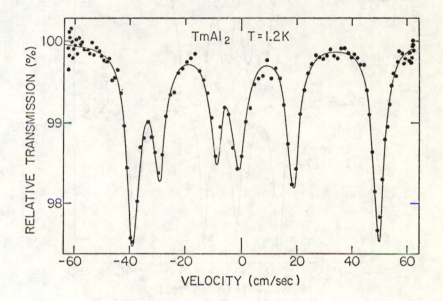

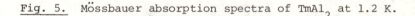

Fig. 5. Mössbauer absorption spectra of $TmAl_2$ at 1.2 K.

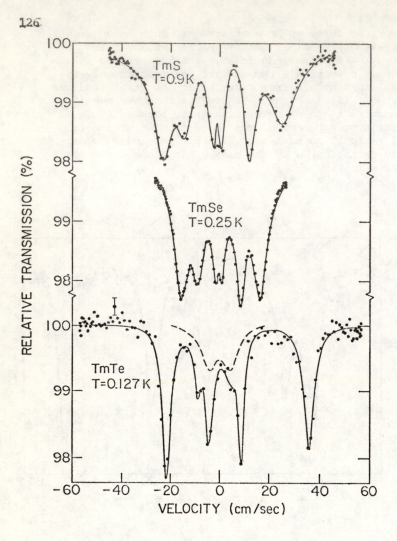

Fig. 6. Mössbauer spectra of TmS, TmSe, and TmTe. The TmS and TmSe materials contain predominantly $Tm^{3+}(4f^{12})$ but TmTe contains predominantly $Tm^{2+}(4f^{13})$.

The calculation discussed earlier for the hyperfine enhanced mechanism was done in the molecular field approximation assuming fast relaxation of the eigenstates of the full Hamiltonian. However the ^{169}Tm nucleus has the fastest characteristic precessional frequencies $\approx 10^{10}$ sec^{-1} of any common Mossbauer transition, and one might question whether fast relaxation (with respect to these nuclear precessional frequencies) will ever be observed in the hyperfine spectra taken with this line. We make this point in Figs. 7 and 8 where we show spectra taken on TmS and TmSe at higher temperatures. These systems are strongly metallic and cubic, and one might suspect they are the most

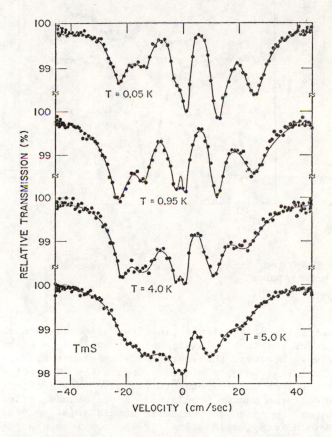

Fig. 7. Selected absorption spectra of TmS taken at higher temperatures showing relaxation-like effects as soon as excited electronic states become populated.

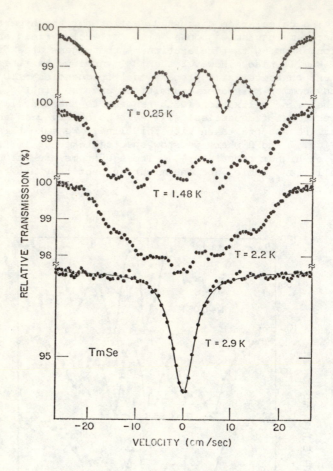

Fig. 8. Selected absorption spectra of TmSe taken at higher temperatures.

likely materials to display collapsing 6-line spectra characteristic of the fast relaxation limit applied to electronic eigenstates. However, the spectra for these materials, like every Tm material we have studied, apparently show relaxation effects when the temperature is raised and excited electronic states become populated. Nevertheless, effects of the hyperfine enhanced mechanism should still be visible and we have no real explanation for the complete absence of the expected effect except to suggest that the molecular field approximation (MFA) may be inadequate in this situation. (The effects of two large terms in the Hamiltonian tend to cancel each other, and the MFA predicts that a much smaller term, the magnetic hyperfine interaction, will control the behavior of the total Hamiltonian.)

Of course, the electronic terms are still dominating the behavior materials ordering above ≈ 0.1 K and, armed with a little bit of knowledge, we are anxious to press into the region of lower transition temperatures where effective nuclear-nuclear interactions play the dominant role. We have already looked at $TmBe_{13}$ in zero magnetic field, but without seeing any ordering to 0.049 K.

I would like to make one additional comment about the importance of Mössbauer experiments in this low temperature area. The attainment of ultralow temperatures is a particularly important means of producing polarized and aligned nuclei for experiments in areas of fundamental interest to physics. The attainment of low temperatures by demagnetization of electronic spin systems has just about been pushed to its logical limit, and hyperfine-enhanced demagnetization of nuclei is expected to play an increasingly important role in the attainment of such ultralow temperatures (a temperature of 1.5 mK after demagnetization has already been reported[4] for $PrTl_3$). The usual mechanism limiting the ultimate temperature attained is simply the hyperfine-enhanced ordering that we have been discussing. Thus an understanding of this phenomena will help to produce a means to polarize or align many different nuclei.

REFERENCES

* This work has been supported in part by the Cottrell Research Corporation and the National Science Foundation.

[1] B. B. Triplett and R. M. White, Phys. Rev. B **7**, 4938 (1973), and references therein.

[2] K. Andres et al., Phys. Rev. Lett. **28**, 1652 (1971), and references therein.

[3] Mössbauer transducer VT-700, Ranger Engineering Corp., Fort Worth, Texas.

[4] K. Andres and E. Bucher, J. Appl. Phys. **42**, 1522 (1971); see also K. Andres and E. Bucher, J. Low Temp. Phys. **9**, 267 (1972).

[5] R. L. Cohen, Phys. Rev. **134**, A94 (1964).

[6] B. D. Dunlap, Mössbauer Effect Methodology, ed. by I. J. Gurverman (Plenum Press, New York, 1971), Vol. 7, p. 123.

MÖSSBAUER MEASUREMENTS AT VERY LOW
TEMPERATURES

L. Bogner, W. Gierisch, W. Potzel, F.J. Litterst and
G.M. Kalvius

Physik Department, Technische Universität München,
D-8046 Garching

In this communication we describe two applications of
very low temperatures to Mössbauer spectroscopy recently
carried out in our laboratory.

a) Biological systems

It has long been recognized, that Mössbauer spectroscopy is a useful tool to investigate the electronic
structure of the iron containing active center of bio-
molecules[1]. In many instances, an essentially static
paramagnetic hyperfine spectrum is observed in such
compounds at low temperatures due to very slow electronic relaxation processes. Measurements are commonly performed in the presence of an external magnetic
field and interpreted with the help of an appropriate
spin Hamiltonian. It is often found in these compounds
that the crystal field splitting of the electronic
ground state is rather small. Even at the temperature
of liquid helium more than one crystal field state is
occupied and the Mössbauer spectra consist of a sum
of several paramagnetic hyperfine pattern. Their relative intensities are determined by the Boltzmann
populations of the different electronic levels. In
order to have only the lowest electronic state populated and to obtain a simple spectrum it is necessary
to reduce the absorber temperature below 1K. A typical
example of such a situation is shown in fig.1 where

the paramagnetic hyperfine spectrum of iron in human transferrin is shown at various temperatures[2]. Transferrin is the iron transport protein of the extracellar fluid. In fig.1 the arrows indicate the posi-

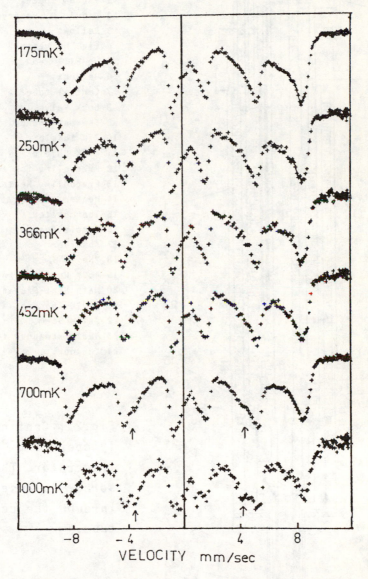

Fig.1: Mössbauer spectra of transferrin powder at 0.025 T longitudinal field and temperatures below 1K.

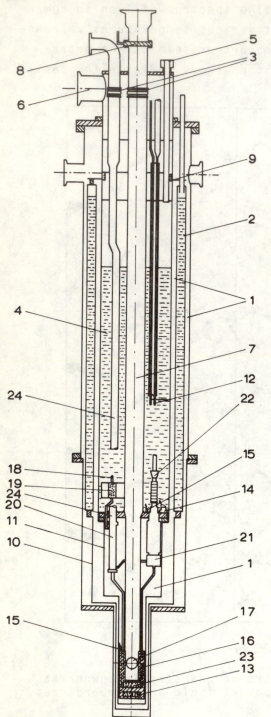

1 Vaccum space
2 Liquid N$_2$ reservoir
3 Bellows
4 Liq. He vessel
5 He Fill
6 He Vent
7 Central tube
8 Lock Valve
9 Thermal Anchor
10 77K Shield
11 4.2K Shield
12 Precooling for sample chamber
13 Precooling Heatexchanger
14 Heater
15 Thermometer
16 Sample Chamber
17 Absorber
18 ^{3}He in
19 4.2K Heat-Exchanger
20 Main Heat-Exchanger
21 Joule-Thomson Valve
22 Valve Stem
23 Heat-Exchanger for sample chamb
24 ^{3}He out

Fig.2a: Continuous-
ly operating ^{3}He re-
frigerator. The ab-
sorber is inserted
through the central
tube.

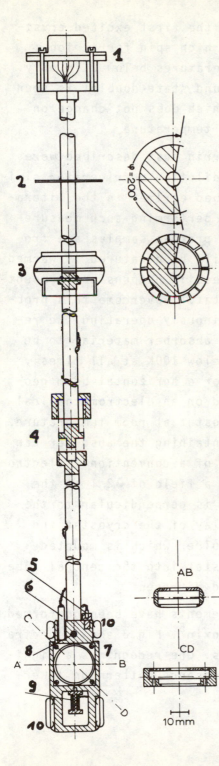

1 Electircal Connector
2 Radiation Shield
3 77K Thermal Anchor
4 Coupling
5 Ge-Thermometer
6 Carbon Thermometer
7 Absorber Holder
8 Heater
9 Heater
10 Contact Springs

Fig.2b: Insert with absorber holder for ³He-refrigerator. The insert can be disconnected at the coupling to be inserted or removed in two parts.

10 mm

tions of lines arising from the first excited cryst-
alline field doublet of the high spin ferric ion.
One recognizes that at temperatures below 0.7K only
the contribution of the ground state doublet is seen
in the Mössbauer spectrum which does not change on
further reduction of sample temperature.

The measurements on transferrin just described were
carried out with a ^{3}He/^{4}He dilution refrigerator similar to an instrument described earlier in the literature[3]. One major problem in performing such measurements is the fact that most protein samples are fro-
zen solutions which often will be denatured if warmed
to room temperature. This easily happens while mount-
ing the sample in the cryostat.To overcome this problem we have designed a continously operating ^{3}He refrigerator which allows the absorber material to be
inserted while being kept below 100K at all times[4].
The cryostat is layed out for a horizontal beam geometry. The source is mounted on an electromechanical
drive motor outside the cryostat at room temperature.
The tail of the cryostat containing the absorber can
be placed between the poles of a conventional electro-
magnet capable of producing a field of ⌣2 T at the
sample. The field direction is perpendicular to the
ɣ-ray. Fig.2a shows the outlay of the cryostat and
fig.2b shows the absorber holder which is mounted
through the top of the cryostat into the central tube
while being cooled by liquid nitrogen.

With this instrument measurements have been performed
on reduced bacterial Ferredoxin[5]. Fig.3 shows spectra
obtained at two temperatures. One recognizes a significant difference in the resonance pattern on lower-
ing the temperature.

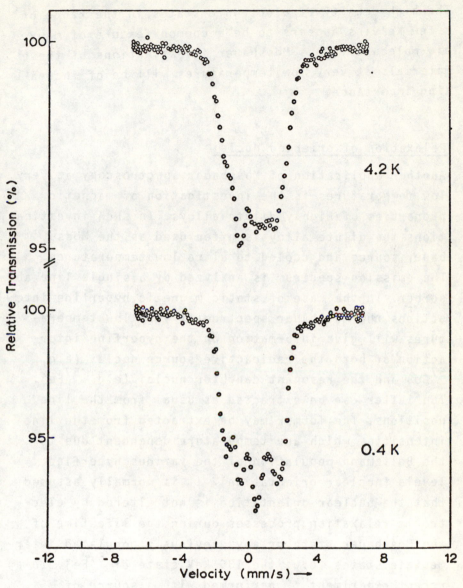

Fig.3: Mössbauer spectra of reduced Ferredoxin in a
transversal magnetic field of 1.4 T.

The high density of electronic states in the range of
a few Kelvins appears to be a common feature of many
bio-molecules. Thus Mössbauer investigations of these
materials at very low temperatures will be of increasing importance.

b) Relaxation of oriented nuclei

Another application of Mössbauer spectroscopy at very
low temperatures is the investigation of magnetic
properties of highly dilute alloys. In such investigations the dilute alloy is often used as the Mössbauer source and cooled to ultra low temperatures.
The emission spectrum is analyzed by a single line absorber. In the case of static magnetic hyperfine interactions the Mössbauer spectrum at very low temperatures will give information on the hyperfine interaction of both the radioactive source nuclei (e.g.
^{57}Co) and the resonant daughter nuclei (e.g. ^{57}Fe).
The latter can be extracted as usual from the line
positions. The former may be extracted from the line
intensities which are temperature dependent due to
the Boltzmann population of the various hyperfine
levels (nuclear orientation). It is normally assumed
that the nuclear orientation is not altered by electronic relaxation processes during the life time of
the Mössbauer state or any previously populated intermediate states (e.g. the 136 keV state of ^{57}Fe). In a
recent experiment[6,7] performed with a source of
~1ppm ^{57}Co in Pd we have found evidence of such relaxation effects. In fig.4 we show spectra of a ^{57}Co in
Pd source at 0.028 mK measured against a ferrocyanide
absorber. Spectrum a) is taken in the absence of an
external field while spectrum b) is recorded with a
field of 1 T in the direction of the γ-ray beam. In

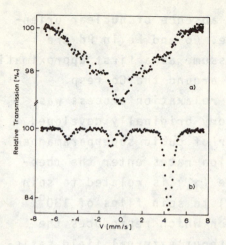

Fig.4: Mössbauer spectra of ^{57}Co in Pd source against a ferrocyanide absorber at 0.028K.
a) $B_{ext}=0$, b) $B_{ext}=1$ T

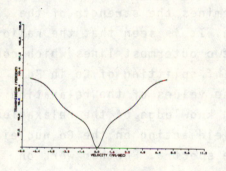

Fig.5: Simulation of relaxation spectrum of ^{57}Co in Pd without external field. The relaxation parameters are $\gamma_\perp=1.3$ MHz and $\gamma_{\parallel}=14.8$MHz.

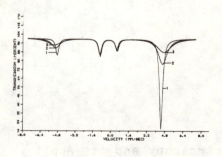

Fig.6: Simulation of relaxation spectra of ^{57}Co in Pd in external field. The parameters are:

	1	2	3
$\gamma_{\parallel}$:	0	0.6	1.2 MHz
$\gamma_\perp$:	0	0.3	0.6 MHz
ß :	0	0.6	1.2 MHz

the zero field spectrum the absence of nuclear orientation is a dominant feature. Co and Fe in Pd are giant moment systems[8]. We assume as a first approximation that the whole giant moment around the Co resp. Fe atom is fluctuating. The relaxation process was described by adapting a theory originally developed for the relaxation behaviour of cubic superparamagnetic particles[9]. Two relaxation rates enter the theoretical description: the one ($\gamma_\perp$) is related to spin flips by 90°, the other ($\gamma_\shortparallel$) to spin flips of 180°. In fig.5 we show a simulation which reproduces the features of the spectrum without external field satisfactorily. In fig.6 some simulations of magnetic patterns in the presence of a magnetic field are shown. The parameter ß determines the strength of the quasimagnetic hf interaction. It is seen that the ratio of line intensities of the two outermost lines which contain the information on the hf splitting of Co in Pd are strongly dependent on the values of the relaxation parameters. Without detailed knowledge of the relaxation processes the magnetic hf field acting on the Co nuclei cannot be extracted from the spectra. Further work on this problem is in progress.

References:

1 G. Lang, in "Mössbauer Spectroscopy and its Applications", Int. Atomic Energy Agency, Vienna STI/PUB/304, 1972, p. 213 ff.

2 C.P. Tsang, L. Bogner and A.J.F. Boyle, Journ. Chem.Phys. 65, 4584 (1976)

3   G.M. Kalvius, T.E. Katila and O.V. Lounasmaa in
    "Mössbauer Effect Methodology", Vol. 5, I.J.
    Gruverman, ed. Plenum Press 1969, p. 231 ff

4   L. Bogner and G.M. Kalvius, Nucl.Instr.Meth. (to be
    published)

5   L. Bogner, PH.D. Thesis, Technical University Munich
    1977 (unpublished); see also: L. Bogner, F. Parak
    and K. Gersonde, Journ. Physique 37 C6-177

6   W. Gierisch, PH.D. Thesis, Technical University
    Munich, 1977 (unpublished)

7   W. Gierisch, W. Koch, F.J. Litterst, G.M. Kalvius
    and P. Steiner, Journ.Magn.Magn.Materials 5, 129
    (1977)

8   G.G. Low and T.M. Holden, Proc.Phys.Soc. 89, 119
    (1966)

9   A.M. Afanas'ev and E.V. Onischenko, Sov.Phys. JETP
    43, 322 (1976)

PANEL DISCUSSION ON USES OF SYNCHROTRON RADIATION

Participants: R. L. Cohen (Leader), P. A. Flinn,
E. Gerdau, J. P. Hannon, S. L. Ruby,
G. T. Trammell

Transcribed from tapes by Eric Shakin, Edited by R. L. Cohen,
and revised by the participants

COHEN: Of the facilities that are available now or will be
operating soon, only a few produce large flux above 10 keV.
The one that I have been working at and that Stan Ruby is
working at is in California, at SSRP (Stanford) - it is a
parasite on the storage ring called SPEAR. The next thing
available is DORIS at Hamburg. DORIS is a large high energy
storage ring which has similar characteristics to SPEAR.
Both of these machines have extremely high circulating
current, on the order of 30-100 ma. They are capable of
> 3 GeV, and they run at rotational periods of about 700
nanoseconds/turn.

GERDAU: In the case of DORIS one can have two possibilities: 480
single bunches with an integral current of about 300 milliamps,
or one bunch with 30-100 ma. The single bunch mode of DORIS
will operate in a few months.

COHEN: Then, there is a synchrotron source in France at Orsay; in
England at Daresbury; both of these don't put out large
amounts of photons above the 5-6 keV regime. At Cornell
there is a synchrotron which is used primarily for high energy
experiments and at Wisconsin there is a storage ring,
TANTALUS, which is useful primarily in the range below 1 keV
photon energy. Being built in the US at Brookhaven is a
large facility dedicated to synchrotron radiation work. It
will have lots of beam pipes and flux. Probably that
machine is 3-4 years off. So, in the US we have funded at
this time as significant projects the Brookhaven machine,
the Wisconsin machine, Cornell (maybe) and Stanford. If you
are in northern Europe, you go to DORIS, if you are in
France, you go to Orsay; and if you are in England, you go
to Daresbury.

QUESTION: Where do you go in the Soviet Union?

COHEN: I don't know anything about the Russian situation.

RUBY: Novosibirsk is apparently up to Vet 3, there has been a
Vet 1 and then Vet 2 and a Vet 3. Budker is a great machine
man and they have done very interesting things in that
laboratory, and the announced figures on the Vet 3 sound
fairly interesting, - they talk about 2.2 GeV max which are
not the most energetic electrons in the world, but the radius

is somewhat smaller than SPEAR, and therefore, the critical
energy will be fairly high. And, they also talk of 100 ma
but no one seems to know much of the details.

COHEN: Shall I put this here for people to look at? This is (writes
reference down) a tabulation by Dr. Gerdau. The reference
is: Optik 45, 1976, 395. That has not only a description
of DESY and DORIS, but also a table which summarizes all of
the current storage ring situations. (Note: Table I has
been added to the transcript of the proceedings.) Now, I
want to get this discussion rolling. Are there questions
that somebody can answer?

CHAPPERT: I have a question about the flux. Yesterday, Dr. Hannon
gave some values for the flux you can expect for a number of
photons per sec per ev per milliradian and he compared it to
the most powerful ^{57}Co source you can get and it looked to
me like these figures were promising -- but after discussing
it with you, I have the feeling that they don't have such
large fluxes.

COHEN: The answer to that apparent discrepancy is that the way to
get these big differences is by considering the solid angle.
There is no way that any synchrotron is going to compete with
a radioactive source within 4π on a photons/eV basis. But,
on the other hand, if you want to do a scattering experiment
or a coherent scattering experiment of some kind where what
you need is extremely well collimated radiation, if you are
talking about 10^{-8} steradians, 10^{-8} is a big factor. The
synchrotron gives you this collimation free because the beam
is 10^{-4} radians high and it is as wide as you want to take
it. You can work easily in solid angles of 10^{-8} steradians and
get big fluxes.

POUND: There are certain possibilities of making lenses however,
which aren't usually considered.

REMARK
BY RUBY: I'd like to make another remark that one of the differences
in Jim's calculations between his top of the page number,
which was the optimistic one, and the so-called real ones
down below, is essentially the very different projections
into the future. For example, one of the points is the word
wigglers. I don't know if this audience knows much about
wigglers. Until now, the particle has been just running
essentially in a simple circle around the synchrotron storage
ring. Suppose running in a simple circle you take an electron
and wiggle it back and forth as it goes around. An extra $\vec{B}$
field that curves it sharply as if it was a small ring and
then bends it back and so you have a wiggling path. This
will give much more light because the radius of curvature is
smaller. It will also accumulate over a long distance

Table I. Synchrotron Radiation Sources
(From BNL 50595, J. P. Blewett, ed., February 1977)
  R = Storage Ring, Y = Synchrotron

Location	Name	Type	Energy (GeV)	Circulating Current (mA)	Status (Operating Constr. or Proposed)
U.S.					
NBS,Washington	SURF I	Y	0.18	5	Op.
Madison,Wisc.	TANTALUS I	R	0.24	40	Op.
Stanford,Cal.	SPEAR	R	4.2	70	Op.
Cornell		Y	12	2	Op.
BNL	NSLS	R	2	1000	Prop
			0.7	1000	Prop
Madison,Wisc.	TANTALUS II	R	2.5	1000	Prop
			0.75	1000	Prop
U.K.					
Daresbury	NINA	Y	5.0	60	Op.
Daresbury	SRS	R	2.0	1000	Prop
France					
Orsay	ACO	R	0.54	35	Op.
Orsay	DCI	R	1.8	400	Constr
Germany					
Bonn		Y	2.5	20	Op.
Hamburg	DESY	Y	7.5	80	Op.
Hamburg	DORIS	R	3.5	900	Op.
Italy					
Frascati		Y	1.1	100	Op.
Frascati	ADONE	R	1.5	60	Op.
Sweden					
Lund		Y	1.2	40	Op.
U.S.S.R.					
Novosibirsk	VEPP 2	Y	0.7		Op.
Novosibirsk	VEPP 3	R	3.5	500	Constr
Moscow	FIAN	Y	0.7		Prop
Moscow	RAKNVA	R	1.3		Constr
Yerevan		Y	6.1	22	Op.
Japan					
Tokyo	INS-SOR	Y	1.3	60	Op.
Tokyo	SOR-RING	R	0.3	100	Op.

because it is not just one bend, but many. Wiggles have been described with enormous increases in flux, but they are difficult to interpret.

COHEN: This flux increase is best projected for the Brookhaven machines. This wiggler that Stan is talking about is this curve here (Fig. 1). You see, it doesn't help you very much at low energies (photon energies below a few keV), but it gives you many orders of magnitude in flux at energies > 10 keV. Then there is another development which is coming on which is called the helical wiggler,* which uses a static magnetic flux that is perpendicular to the electron path, but whose direction rotates as you go along the path. The helical wiggler allows you, for the first time, to get synchrotron radiation which is semi-tuned rather than being broad band, depending on the particular field configuration that you pick. The straight wiggler, the kind of wiggler that Stan was talking about, has alternating regions of field down and field up. As the electron goes along it gets banged from side to side. This helical wiggler is technologically more difficult to make because it requires very short sections with large $\vec{B}$ fields. The electron goes in this direction (hand motions). What it sees is a $\vec{B}$ field that rotates this way and as the e- is accelerated in this kind of pattern, it puts out radiation in the forward direction with predominant photon energy determined essentially by the pitch of the helix. So for the very high energies above 10-20 keV, this helical wiggler is going to be enormously effective at giving you a lot of flux. It puts all the energy in the high energy part.

GERDAU: Can you remark on the angular distribution of the radiation coming from the wiggler?

COHEN: I don't know that. If you do, please take the pen.

RUBY: I'll just make a modest response to that. If you do strong wiggling then you will actually change the electron velocity vector a little to the right and a little to the left. This inreases the beam divergence and for some special experiments that require the most meticulous handling of the photons (like nuclear Braggscattering), this disadvantage may be important. Clearly the wigglers will help a lot. For our experiment, perhaps less than for others.

COHEN: There are details in the wiggler that are important.

TRAMMELL: We have smaller angles than we really need, so let's not knock down the wiggler. If the angle should open up to 10^{-3}, it's ok.

*B. M. Kincaid, J. Appl. Phys. 48, 2684 (1977).

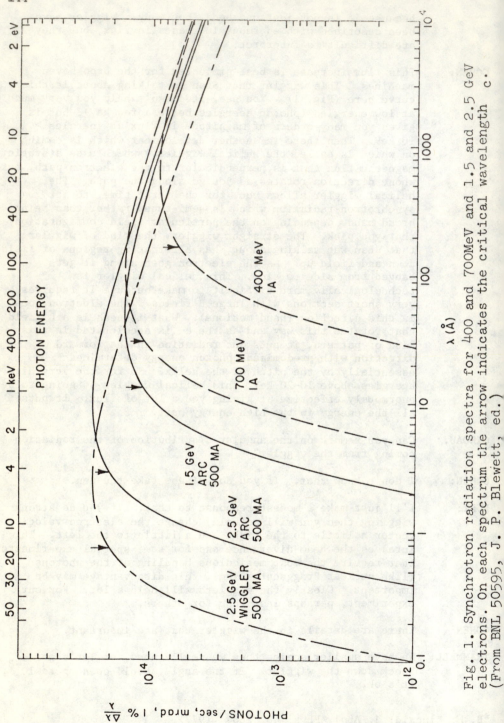

Fig. 1. Synchrotron radiation spectra for 400 and 700MeV and 1.5 and 2.5 GeV electrons. On each spectrum the arrow indicates the critical wavelength c. (From BNL 50595, J. P. Blewett, ed.)

RUBY: With your thin crystal, Jim means that if we can make nuclear mirrors thin for x-ray scattering (while thick for nuclear) then some of the angular effects are not so severe.

POUND: These things are sort of interactive in the sense that the simple alternating wiggler essentially translates that single pulse one sees (when the forward beam goes through) into a sequence of pulses each time it goes through. So that is the simple gain. And, the length of the pulse spread is shorter because it's bending faster. That's why it spreads to higher energies.

PFEIFFER: Do we have to wait until Brookhaven before we get a wiggler?

COHEN: The people at SSRP would, of course, like to put wigglers on immediately. The people who own the storage ring are not very enthusiastic about letting people play with the beam orbit that way. It has not yet been demonstrated that you can put these things on and maintain the stability of the electron orbits.

POUND: There was a wiggler in the CEA machine which was there for different reasons. It was there to provide radiation damping of orbit oscillations in the storage mode of the machine, and therefore we know that the wiggler is a useable thing in conjunction with a synchrotron.

PFEIFFER: It occurs to me if you got up to 100 keV which the chart seems to say you can do in energy, you would still have a flux of 10^{12}. Then you might be able to try some sort of a pumping experiment where you would pump some level which was not very sharp, which was in fact super broad, but higher in energy. Once populated this level would cascade down to the level of interest. That way you can get some pumping gain from the fact that you pump a fast broad lifetime and it feeds the metastable state. I think this becomes a possibility one can consider with a wiggler.

RUBY: That's right. You have a combination of the virtues of a fast level getting a lot of flux. It is an interesting idea; as far as I know, it is the first time I've heard it.

COHEN: If you really want to go to high energies, what you want is SPEAR, because SPEAR has the higher electron energy. SPEAR sometimes runs at energies where 100 keV is not out of the question. They are also talking about putting wigglers on SPEAR.

KALVIUS: Is it of some advantage if you want to do scattering experiments if you have defined angles of synchrotron radiation. Now most scattering experiments need a sharp angle definition in one direction, but not in the other, so you can open up, for instance, a Bragg reflection to 1° or 2° and still get reasonably good solid angle. Then, of course, many advantages of the synchrotron radiation against a radioactive source begin to evaporate. That's my question: How does it look if you compare the source and open up the solid angle in one direction which is allowed for most applications?

TRAMMELL: Yes, we can find it out in one direction. We are at 10^{-4} in the plane of the orbit. For most scattering experiments that is way too high collimation. You don't need that accurate a collimation. You don't want it for any absorption experiment. No one has taken advantage of the fact that, say for the ^{57}Fe line, you can have an interference experiment. Or, as you say, microscopes, lenses, etc. with various rays coming with different optical path lengths could be made with optical differences up to 30 m. This has never been exploited and seems to me would be a very good possibility. Another thing that we should say concerning this, is that at this 1 Å level, we have a wavetrain that is 30 m long. If you go to x-rays, you have a wavepacket 1 micron long. The Mössbauer will see its 30 m and you would have a little microscope to see the wavepacket of the x-ray.

COHEN: That's talking about the coherence length once you have gone through a Mössbauer-type filter.

TRAMMELL: Eventually, we may have an interferometer (draws on viewgraph) "here comes your beam in, you've got a crystal here, and you split it into two beams." You are later going to bring them back together and observe fringes and what you want is to do something on one leg and see this effect of advancing this phase by π. With this thing that we have which is very narrow (you have to come in with a Bragg angle of 10^{-4}) you need that in order to make this business to get these two waves in good phase. We've got 30 m to play with. We could make this path length 30 m longer than the other. That opens up this possibility. That is, one of these legs we actually make it 30 m, but we do it on the table, viz, we have a couple of crystals very accurately placed here and also 180° scattering. In other words, a long grating and we double the path here. We finally let it come back out. You can use a monolayer of Fe and maybe send this beam through it a thousand times, and get a lower Mössbauer signal. From that you have to have a very well collimated beam because the width of these Bragg reflections are about 10^{-4}. So, it has got to be that well collimated.

QUESTION: What determines the coherence length?

RUBY: The coherence length $L = \lambda(\nu/\Delta\nu) \approx 10^2$ meters for ^{57}Fe when $\Delta\nu/\nu = 10^{-12}$ as it is with normal radioactive decay of a Co parent. However, when you reflect white light off nuclear mirrors there can be 'enhancement of the radiative channel.' In our case, this can be about a factor of 10 and the emitted light will have $\Delta\nu/\nu \approx 10^{-11}$, along with a lifetime of about 10 nanoseconds. This broadened source is disadvantageous for experiments involving high resolution conventional Mössbauer spectroscopy. The enchancement, however, multiplies the number of reflected photons by about ten over simple predictions, an advantage for interferometry, etc.

TRAMMELL: And also, you have 10 m, that is quite a long coherence length.

RUBY: More than needed; even 1 m is quite acceptable.

TRAMMELL: It seems to be that this long coherence length opens up a fantastic possibility here. If you move these things 10^{-12} apart you will change the fringe by π.

POUND: We already have those problems in optical interferometers at that level. The problem in this device is with dimensional stability.

TRAMMELL: Yes, that would have $\Delta\lambda/\lambda$ capability you mean.

PFEIFFER: If you make 1000 bounces, don't you also have some sort of problem with recoil free fraction?

TRAMMELL: No, we think that we can get maybe 70% reflectivity at each, or 50%, let's say, within a factor of 2.

GERDAU: I think that we are one step too far. We are speaking about beautiful experiments which can be done if one has a cleaned beam. I think one should make some remarks on what are the conditions to see a beam. If you agree, I'll give you some numbers which as I think are important if one wants to see an effect. The center of the beam of γ-rays coming out is tangent to the electron orbit. This is the main source of the horizontal divergence. Furthermore you have the vertical divergence determined by the radiation characteristics of the synchrotron radiation. At an electron energy of 3 GeV the vertical divergence is about $\pm$ 15". Superimposed is the angular divergence of the electron orbit. This divergence adds to the diyergence coming from the synchrotron radiation. So if one takes the numbers of DORIS one gets a vertical divergence of $\pm$ 30". The next question is how broad are the Bragg-reflexes in the system. If one takes Ge or Si and looks at 90° reflections one starts with widths of about 1". So one loses about a factor of 60 at the beginning, and all these beautiful numbers mentioned will not work. From my point of view, the first question one has to ask is: If I take a reflecting crystal which contains iron, how broad are the reflexes? At exact Bragg angle one probably has reflectivity in a broad energy range (~100 Γ or more). So even if the vertical angular width is small one may get enough intensity but one observes only an effect of at most 1% or less because one looks with a "sharp" absorber at a broad distribution. Yesterday in his talk Professor Hannon showed what can be done. If one regards the reflectivity drawn in an angle-energy diagram one sees that if one looks for example at the 50% reflectivity line that reflectivity in a small energy range (for example 2 Γ) is obtained if one goes a bit off the Bragg angle. But the angular width of this region is small. And again one has a serious intensity reduction. So, if one wants to do a real experiment one should think first about these experimental conditions and various other factors of reduction one gets.

FLINN: We have talked some about these problems of angular spread; there are some well known techniques which enable one to increase considerably the acceptance angle given by the analysis. The most useful if the asymmetric cut of the monochromator crystal. The natural thing is to put a crystal in with the crystal surface parallel to the reflecting planes, but instead you cut the crystal so that the physical surface is at a very small angle to the beam. Then the acceptance angle is increased by

$$\sqrt{\frac{\text{Sin (Angle In)}}{\text{Sin (Angle Out)}}}$$

So, if you get large crystals and orient them accurately, you can pick up an order of magnitude in intensity this way. A further increase can be obtained by using a higher Z element in place of Ge or Si. If you go to one of the systems other than polarization, such as using a crystal made with Fe and pseudo-Fe, so the electronic scattering cancels, then you do not need the monochromator. You put your Fe alloy crystal directly in the beam and you can use the full horizontal divergence and get probably 20" of vertical rather than the 1 or 2 seconds.

LIST OF ATTENDEES

Mark Anderson
The Johns Hopkins University
Baltimore, Maryland 21218

Salil Banerjee
Physics Department
State University of New York
Stony Brook, New York 11794

Michael R. Blizzard
Physics Department
University of Cincinnati
Cincinnati, Ohio 45221

Ralph H. Castain
Physics Department
Purdue University
West Lafayette, Indiana 47906

David C. Champeney
School of Mathematics & Physics
University of East Anglia
Norwich, Norfolk, NR4 7TJ
United Kingdom

Jacques Chappert
Centre d'Etudes Nucleaires
85X - 38041 Grenoble - Cedex
France

Richard L. Cohen
Bell Laboratories
Murray Hill, New Jersey 07974

Solly G. Cohen
Racah Institute of Physics
Hebrew University
Jerusalem
Israel

Romain Coussement
Instituut voor Kern & Stralingsfysika
Dept. Natuurkunde
University of Leuven
Celestijnenlaan 200 D
B-3030 Heverlee
Belgium

James M. Daniels
Department of Physics
University of Toronto
Toronto, Ontario
Canada M5S 1A7

Peter G. Debrunner
Physics Department
University of Illinois
Urbana, Illinois 61801

Hendrik De Waard
Physics Department
The University
Westersingel 34
Groningen
The Netherlands

Bobby D. Dunlap
Solid State Science Division
Argonne National Laboratory
9700 South Cass Avenue
Argonne, Illinois 60439

Marcu Eibschutz
Room 10-431
Bell Laboratories
Murray Hill, New Jersey 07974

Paul A. Flinn
Physics Division
Argonne National Laboratory
9700 South Cass Avenue
Argonne, Illinois 60439

Frank Y. Fradin
Materials Science Division
Argonne National Laboratory
9700 South Cass Avenue
Argonne, Illinois 60439

Gerald T. Garvey
Physics Division
Argonne National Laboratory
9700 South Cass Avenue
Argonne, Illinois 60439

150

Erich Gerdau
Universität Hamburg
Lutuper Chaussee 149
2000 Hamburg 50
Germany

Ulrich Gonser
Universität des Saarlandes
Fachbereich Angewandte Physik
6600 Saarbrücken
Germany

Stanley S. Hanna
Max-Planck Institut für Kernphysik
Postfach 103980
6900 Heidelberg 1
West Germany

James P. Hannon
Physics Department
Rice University
Houston, Texas 77001

Gilbert R. Hoy
Physics Department
Boston University
Boston, Massachusetts 02215

Günter Kaindl
Institut für Atom-u. Festkörperphysik
Freie Universität Berlin
Boltzmannstrasse 20
1000 Berlin 331
Germany

Georg M. Kalvius
Physik Department E15
Techn. Univ. München
D8046 Garching
James Franck Str.
Germany

Clyde W. Kimball
Materials Science Division
Argonne National Laboratory
9700 South Cass Avenue
Argonne, Illinois 60439

Ottmar C. Kistner
Physics Department
Brookhaven National Laboratory
Building 510
Upton, New York 11973

Noémie Koller
Department of Physics
Rutgers University
New Brunswick, New Jersey 0890

Walter Kündig
Physics Institute
University of Zurich
Schonberggasse 9
8001 Zurich
Switzerland

Guido F. Langouche
Instituut voor Kern- en
    Stralingsfysika
Leuven University
Physics Department
Celestijnen laan
B-3030 Leuven
Belgium

Mou Ching Lin
Department of Physics
Northern Illinois University
DeKalb, Illinois 60115

Harry J. Lipkin
Physics Division
Argonne National Laboratory
9700 South Cass Avenue
Argonne, Illinois 60439

John C. Love
Physics Division
Florida Institute of
    Technology
305 Arcadia Court
Melbourne, Florida 32901

James G. Mullen
Physics Department
Purdue University
West Lafayette, Indiana 47906

Felix E. Obenshain
Oak Ridge National Laboratory
P. O. Box X, Building 6003
Oak Ridge, Tennessee 37830

Shimon Ofer
Racah Institute of Physics
The Hebrew University
Jerusalem
Israel

Moshe Pasternak
Physics Department
University of California
Santa Barbara, California 93106

Gilbert J. Perlow
Physics Division
Argonne National Laboratory
9700 South Cass Avenue
Argonne, Illinois 60439

Loren N. Pfeiffer
Bell Laboratories
Murray Hill, New Jersey 07974

Walter A. Potzel
Physik Department E15
Technische Universität München
D-8046 Garching, München
Germany

R. V. Pound
Physics Department
Harvard University
Lyman Laboratory of Physics
Cambridge, Massachusetts 02138

Richard S. Preston
Physics Department
Northern Illinois University
DeKalb, Illinois 60115

Gottipaty N. Rao
Physics Department
Indian Institute of Technology
Kanpur-208016
India

R. L. Rasera
Department of Physics
University of Maryland -
  Baltimore County
5401 Wilkens Avenue
Baltimore, Maryland 21228

Sipke R. Reintsema
Instituut voor Kern- &
  Stralingsfysika
Dept. Natuurkunde I.K.S.
Celestijnenlaan 200 D
B-3030 Heverlee
Belgium

Stanley L. Ruby
Physics Division
Argonne National Laboratory
9700 South Cass Avenue
Argonne, Illinois 60439

Lyle H. Schwartz
Materials Science & Engineering
  Department
Northwestern University
Evanston, Illinois 60201

Gopal K. Shenoy
Solid State Science Division
Argonne National Laboratory
9700 South Cass Avenue
Argonne, Illinois 60439

James R. Stevenson
Physics Department
Purdue University
West Lafayette, Indiana 47907

R. Dean Taylor
Los Alamos Scientific Laboratory
P. O. Box 1663, MS-764
Los Alamos, New Mexico 87545

Michael A. Tenhouer
Department of Physics
University of Cincinnati
Cincinnati, Ohio 45221

152

George T. Trammell
Physics Department
Rice University
Houston, Texas 77001

Baylor B. Triplett
Department of Physics
Stanford University
Stanford, California 94305

Jan M. Trooster
University of Nymegen
Molecuul Spectroscopy
Faeulkit Wes - en Natuurkunde
Toernooiveld, Nijmegen
Holland

Marc Van Rossum
Instituut voor Kern- en
    Stralingsfysica
University of Leuven
I.K.S., Dept. Natuurkunde
Celestijnenlaan 200 D
3030 Leuven
Belgium

William T. Vetterling
Department of Physics
Harvard University
Lyman Laboratory
Cambridge, Massachusetts 02138

Friedrich E. Wagner
Physik Department E15
Technical University of
    Munich
D-8046 Garching
Germany

J. C. Walker
Department of Physics
The Johns Hopkins University
Baltimore, Maryland 21218

Roger Wäppling
Institute of Physics
Box 530
S751 21 Uppsala
Sweden

Heiner Winkler
Department of Physics
University of Illinois
Urbana, Illinois 61801

Gerhard H. Wortmann
Physics Division
Argonne National Laboratory
9700 South Cass Avenue
Argonne, Illinois 60439